The Golden Ratio

Φ = 1.618...

The Facts and the Myths

Francis D. Hauser, PhD

The Golden Ratio: The Facts and the Myths by Francis D. Hauser

Copyright © 2015 Francis D. Hauser. All rights reserved.

ISBN-13: 978-1517518776
ISBN-10: 1517518776

Library of Congress Number: 2015915899
CreateSpace Independent Publishing Platform, North Charleston, SC

Synopsis

Euclid, a Greek mathematician, wrote *The Elements* twenty-three hundred years ago. This book is regarded as history's oldest and still referenced textbook. It is primarily about geometry and contains dozens of figures. Five of these are constructed using a line that Euclid says "is cut in extreme and mean ratio." Today this is called the golden ratio and often referred to with the symbol Φ.

Many myths have grown up around this ratio. For example, Kepler, of solar system fame, thought that Φ was "a tool of God." I wrote this book to learn about these myths.

From *The Elements*, I learned how Euclid used Φ. I read about the myths. They arise from very famous people such as Pythagoras, Plato, Pacioli, Kepler, and Fibonacci. I found there is a common thread among the myths. Φ is an irrational number (a number whose digits go on forever and never form a repeating pattern). Φ can be used to draw pleasing figures using only an ungraded straightedge and compass. But because it is irrational, its numerical value cannot be written down using integers and fractions, which were the only numbers used in Euclid's time.

Mathematicians before Euclid knew that irrational numbers existed. They could describe them with words and figures. But to many people, a number unable to be written down was absurd.

Because of Euclid's book, the existence of Φ became widely known. For centuries, when there was a close connection between religion, science, and math, even great scientists and engineers believed that Φ was godlike.

This book discusses the myths from an engineering viewpoint. Were the pyramids built using Φ? Is the golden spiral really a logarithmic spiral? Are the Fibonacci numbers patterned after Φ?

The last chapter of the book shows how Euclid handled irrational numbers; how Euclid did algebra using geometry; and a simple visual proof of why there are only five Platonic solids.

About the Author

Francis D. Hauser earned his PhD in electrical engineering from the University of Denver in 1972.

He is a dynamic systems analyst in the motions and control of launch and reentry vehicles, multibody spacecraft, fixed and rotor-wing aircraft, large high-speed oceangoing watercraft, landcraft, and wind-driven turbines.

He is a university and college lecturer in graduate, undergraduate, and continuing-education courses on general optimization theory, Newtonian mechanics (statics and dynamics), linear algebra, Fourier analysis, and conventional, modern, and space-vehicle control.

The figures in this book were drawn using methods in Dr. Hauser's other book: *Excel with VBA for Engineers and Mathematicians*, CreateSpace Independent Publishing, 2015. The application, Excel with VBA, resides in Microsoft Office.

Acknowledgment

Several years ago, my friend Melco Absin handed me a note that led to this book.
Do you know anything about this number, $\Phi = 1.6180339887\ldots?$

Abraham Lincoln Quoted Euclid

Things that are equal to the same thing are also equal to each other.
This is the First Common Notion in Book I of *The Elements*.

About the Cover

The Elements is a geometry textbook that is still valid after twenty-three hundred years. It was written by Euclid, a Greek mathematician. It contains 465 theorems (propositions). One of them concerns a line segmented into lengths that are in *golden ratio* to one another. The following figure shows a line of length (Φ+1), with segment lengths of Φ and unity.

The ratio between these lengths is: $\dfrac{\Phi+1}{\Phi} = \dfrac{\Phi}{1} = \Phi$

Φ has come to be called *the golden ratio*. In *The Elements*, the following five figures are constructed. The construction of each is fundamentally based on Φ.

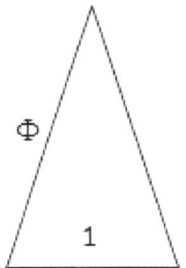

Isosceles triangle whose base angles are twice its apex angle.

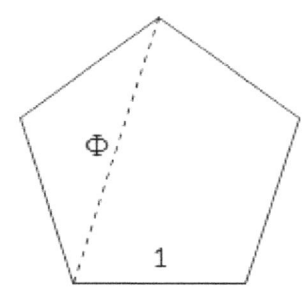

The regular pentagon, which has five equal sides inscribed in a circle.

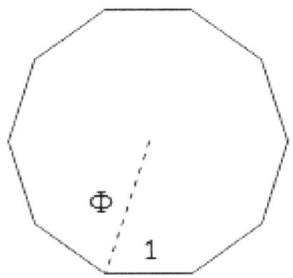

The regular decagon, which has ten equal sides inscribed in a circle.

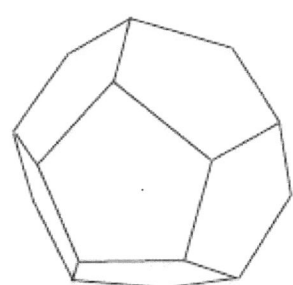

The regular dodecahedron, which has twelve equal regular-pentagon faces inscribed in a sphere. This is one of the five Platonic solids.

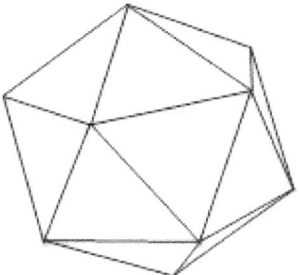

The regular icosahedron, which has twenty equal regular-triangle faces inscribed in a sphere. This is one of the five Platonic solids.

Table of Contents

Introduction

Euclid wrote *The Elements* twenty-three hundred years ago. It is regarded as history's oldest math textbook that is still valid and still referenced. It includes virtually all of the math that was known at the time, but is principally about geometry.

The earliest actual copy is a Greek edition by Theon, published sixteen hundred years ago. Historians think that parts of the manuscript were written by others, including Pythagoras, Eudoxus, and Thaetetus. Euclid collected these into a single manuscript along with his own work.

Euclid constructs more than 465 figures in *The Elements*, at least one for each of its 465 theorems (propositions). Five of these figures are constructed using a geometric quantity that mythology has come to call the "golden ratio." Euclid doesn't use that name. Instead, he calls it the "extreme and mean ratio," and defines it as follows: "A straight line is said to have been cut in extreme and mean ratio when, as the whole line is to the greater segment, so is the greater to the less."

Let's define the whole line as *L,* the greater segment as *a*, and the lesser as *b*. Now, if we rewrite this proposition using math notation: "*L* is said to have been cut in extreme and mean ratio when, as *L* is to *a*, so is *a* to *b*."

Referring to figure 1, L = a + b. When $\dfrac{L}{a} = \dfrac{a}{b}$, L has been cut in extreme and mean ratio.

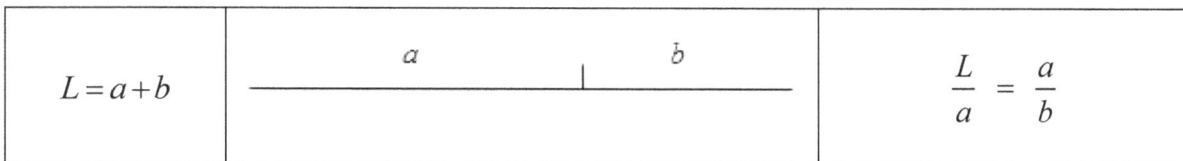

Figure 1: A Line Cut in the Extreme and Mean Ratio

Therefore, either $\dfrac{L}{a}$ or $\dfrac{a}{b}$ is the extreme and mean ratio, a.k.a. the golden ratio.

Mythology has also come to use the symbol Φ when referring to the golden ratio. It will be shown later that the value of Φ is approximately 1.618. Euclid himself was unable to calculate or write down this value because of the limited knowledge of mathematics that existed at the time.

The following table shows the five figures whose construction is fundamentally based on the golden ratio.

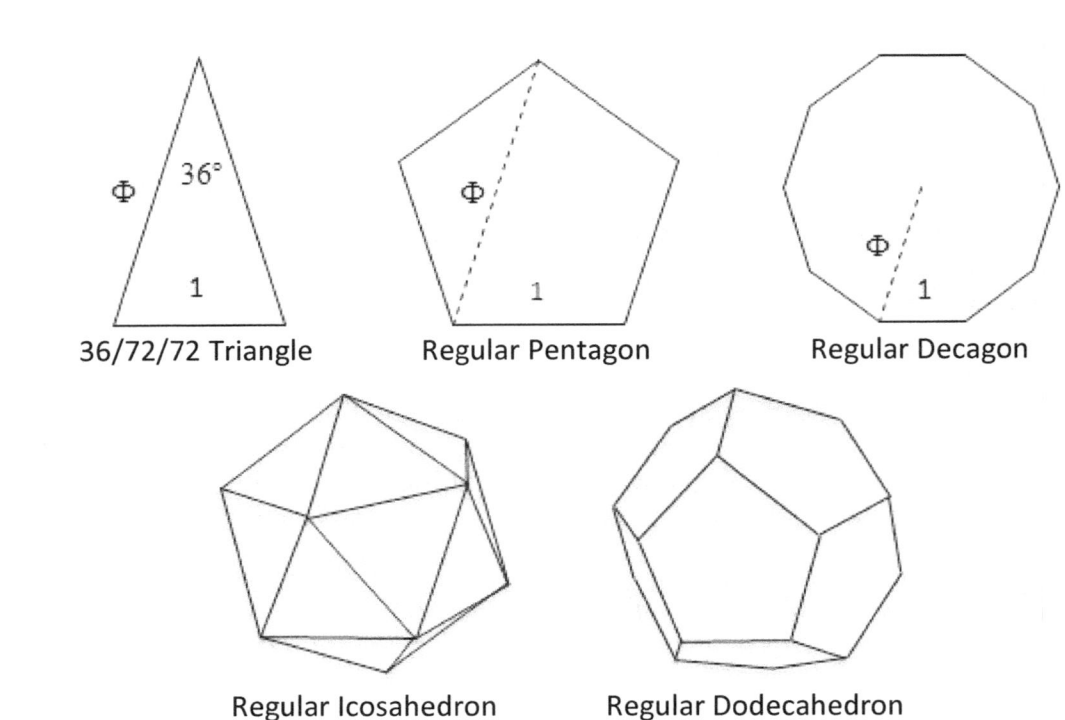

36/72/72 Triangle   **Regular Pentagon**   **Regular Decagon**

Regular Icosahedron   **Regular Dodecahedron**

• *36/72/72 Isosceles Triangle.* The ratio of its side to its base is Φ. This is the same as having each base angle (72°) double the remaining one (36°).

• *Regular Pentagon.* This figure has five equal sides inscribed in a circle. This gives it a diagonal-to-side ratio of Φ.

• *Regular Decagon.* This figure has ten equal sides inscribed in a circle. The ratio of the radius of this circle to each of the sides is Φ. Hence, each side is the base of a 36/72/72 triangle.

• *Regular Icosahedron*. This three-dimensional figure is bounded by twenty equal faces inscribed in a sphere. Each face is a regular triangle, and Φ is used to build its internal scaffolding. It is one of the five Platonic solids.

• *Regular Dodecahedron*. This three-dimensional figure is bounded by twelve equal faces inscribed in a sphere. Each face is a regular pentagon, and Φ is used to build its internal scaffolding. It is one of the Platonic solids.

Figure 2: Figures Constructed Using the Golden Ratio

Euclid never used the name *the golden ratio*, and he never used its symbol or its value. The mythological name was not used until the Renaissance, and that was eighteen hundred years after Euclid. Also, Euclid never mentions mysticism or divinity in his strictly math-based textbook. However, in this treatise, the name *the golden ratio* and its symbol Φ will be used.

The purpose of this treatise is to:

1. Define Φ in Euclid's words.

2. Discover how Euclid used it in *The Elements*.

3. Study the principal myths that have grown up around Φ.

The material in this treatise came from all the books listed in the bibliography. The main source for the facts about Φ is the translation of *The Elements* by Sir Thomas Heath.[1] When a quote is from *The Elements*, it is from Heath. The main sources for the myths about Φ are Mario Livio[2] and Peter Bentley.[3]

Notes about *The Elements*

1. When *The Elements* was written, math consisted only of counting, arithmetic, and geometry. Algebra was invented about six hundred years after Euclid.

2. Euclid wrote *The Elements* using word problems, word solutions, and figures. He did not use equations. But algebraic equations will be used in this treatise for ease of understanding when illustrating Euclid's methods. Often, Euclid's geometric method will be shown together with the modern algebraic methods.

3. In Euclid's time, only positive integer numbers were used. Negative numbers were handled as positive debt. Fractions, which are ratios of integers, were used. Mathematicians knew that irrational numbers existed, but they could not write their values since there was no decimal point. This is why Euclid never computed the value of Φ. He could only define and use it in words and figures.

4. There were no measurement standards. Everybody used different lengths and different angle measures. The circle had not yet been divided into 360 degrees. When the angle was 90 degrees, Euclid called it a right angle. He used the right-angle-and-a-fifth for 108 degrees. Euclid referred to the 36/72/72 triangle as an isosceles triangle having its base angles double the remaining one.

5. Euclid's tools were an unmarked straightedge and an unmarked collapsible compass. What we call algebra today, Euclid did using geometry. This will be shown extensively throughout this book.

6. In Euclid's time, books were manuscripts on papyrus and were copied by hand. This is why geometry was so important. *A picture is worth a thousand words*.

7. Euclid's *Elements* included virtually all math known at the time. It remained practically unaltered for two thousand years. Practical geometry today is known as *Euclidean geometry*.

---

1. Sir Thomas Heath, *Euclid, the Thirteen Books of the Elements* (New York: Dover Publications, 1956).
2. Mario Livio, *The Golden Ratio* (New York: Broadway Books, 2002).
3. Peter J. Bentley, *The Book of Numbers* (Buffalo, NY: Firefly Books, 2008).

Summaries of the Chapters and the Appendix

1. The Facts about Φ: In this chapter, Φ is defined, and the five figures that are based on Φ are drawn. Just as Euclid would have done it, these figures were drawn using only an unmarked straightedge and a collapsible compass without degree markings.

2. The Myths about Φ: The principal myths about Φ are centuries old and are associated with important historical figures. The discussions in this chapter are supported by mathematical analysis.

For example, the dimensions of the large Egyptian pyramids are compared to Φ. Patterns in nature, such as snail shells, resemble a spiral that can be drawn using Φ. This so-called golden spiral is compared to the mathematical spiral that is called logarithmic. The Fibonacci number series is associated with Φ. This series is analyzed using the mathematics of recursive series.

3. Book XIII and Φ: *The Elements* is divided into books. Theorems and construction methods are called propositions. *Book XIII* is entirely about the Platonic solids.

This chapter covers all of the propositions in *Book XIII* that involve Φ. The reader can see how Euclid did algebra using geometry. The reader can see how Euclid handled irrational numbers. In Euclid's time, the only numbers that existed were positive integers. Euclid could not write down the values of irrational numbers, so he gave them names and described them using phrases such as "the ratio of the greater segment to the lesser segment."

This chapter also discusses the famous proposition 18, which proves that there can only be five Platonic solids. In other words, there can only be five regular polyhedrons, which are 3-D figures contained entirely by equal straight lines that form equal faces.

Appendix: It contains a history and proof of the Pythagorean theorem.

Chapter 1: The Facts about Φ

The material in this chapter defines Φ and constructs the five figures whose foundations are based on this ratio. An effort has been made to construct these figures the way Euclid did (i.e., using only a straightedge and compass, both ungraded). The following is the outline of this chapter.

1.1 The Arithmetic Definition of Φ

- $\dfrac{L}{a} = \dfrac{a}{b}$ where $L = a + b$
- Definition 3 in Book VI of *The Elements*
- A line that is cut in Φ proportion
- The numerical value of Φ
- Ways of representing a line that is cut in Φ proportion

1.2 The Geometric Definition of Φ

- $a^2 = Lb$ where $L = a + b$
- Proposition 11 in Book II of *The Elements*
- Comparing the area a^2 to the area Lb
- Relating the arithmetic and geometric definitions of Φ

1.3 How to Cut a Line in Φ Proportion

- Proposition 30 in Book VI of *The Elements*
- Cutting a line in Φ proportion
- Algebraic proof of this method

1.4 The 36/72/72 Triangle and the Regular Decagon

- Proposition 10 in Book IV of *The Elements*
- Constructing these figures
- Algebraic proof of their interior angles

1.5 The Regular Pentagon

- Proposition 11 in Book IV of *The Elements*
- Constructing this figure

1.6 The Platonic Solids

- Defined and illustrated

1.7 The Golden Rectangle

- Defined and illustrated

1.8 The Regular Icosahedron

- Proposition 16 in Book XIII of *The Elements*
- Constructing this figure

1.9 The Regular Dodecahedron

- Proposition 17 in Book XIII of *The Elements*
- Constructing this figure

1.10 Modern-Day Arithmetic Properties of Φ

- Ways of representing Φ that were not possible in Euclid's time

1.1 The Arithmetic Definition of Φ

$$\frac{L}{a} = \frac{a}{b} \quad \text{where} \quad L = a + b$$

Definition 3 in Book VI of *The Elements* is:

"A straight line is said to have been cut in extreme and mean ratio when, as the whole line is to the greater segment, so is the greater to the less."

In the following figure, the whole line is *L*, the greater segment is *a*, and the less is *b*.

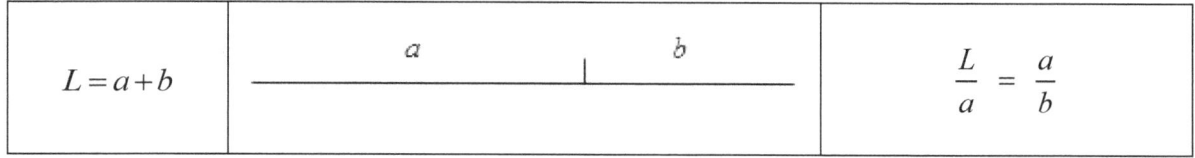

With $L = a + b$, (L/a) is what Euclid called "the extreme and mean ratio." Today, this has come to be called the golden ratio. It is also often indicated by the symbol Φ. It is a special way to proportion a line. Euclid used this proportion to construct five figures, and these were shown in figure 2.

Due to the math available, Euclid could not calculate the numerical value of Φ. Since its value will make it more understandable, we will now calculate it using algebra.

We will solve the equation: $\dfrac{L}{a} = \dfrac{a}{b}$ where $L = a + b$:

Since $(b = L - a)$: $\dfrac{L}{a} = \dfrac{a}{L-a}$

Cross multiplying: $L(L - a) = a^2$

Rearranging: $\left(\dfrac{L}{a}\right)^2 - \left(\dfrac{L}{a}\right) - 1 = 0$

Finally, substituting $\Phi = (L/a)$: $\Phi^2 - \Phi - 1 = 0$

This is an ordinary algebraic equation. The solution is given by the familiar quadratic formula, taught in virtually every elementary school in the world. It has two solutions.

The positive solution is: $\Phi = \dfrac{1 + \sqrt{5}}{2} = 1.6180339887...$

The negative solution is: $\Phi = \dfrac{1 - \sqrt{5}}{2} = -0.6180339887...$

Since Φ is a ratio of lengths, only the positive solution makes sense. Note that the digits of Φ never form a repeating pattern as they go on and on. That means that Φ is an irrational number, and can't be represented using integers and fractions. This is why Euclid could not compute its value. He had to refer to it in a ratio with a known number.

As shown in the following figure, Euclid's line can appear in other forms.

$L = a+b$	$\begin{array}{cc} a & b \\ \rule{3cm}{0.4pt} \end{array}$	$\dfrac{L}{a} = \dfrac{a}{b} = \Phi$
$L = \Phi+1$	$\begin{array}{cc} \Phi & 1 \\ \rule{3cm}{0.4pt} \end{array}$	$\dfrac{\Phi+1}{\Phi} = \dfrac{\Phi}{1} = \Phi$
$L = \Phi+1+\Phi$	$\begin{array}{cc} \Phi+1 & \Phi \\ \rule{3cm}{0.4pt} \end{array}$	$\dfrac{\Phi+1+\Phi}{\Phi+1} = \dfrac{\Phi+1}{\Phi} = \Phi$
$L = 1$	$\begin{array}{cc} .618\ldots & .382\ldots \\ \rule{3cm}{0.4pt} \end{array}$	$\dfrac{.618+.382}{.618} = \dfrac{.618}{.382} = \Phi$

Figure 3: Euclid's Line

Figure 3 further illustrates how Euclid avoids computing the value of Φ. There are three lengths involved, and one of them is a known (rational) number. For example, if $b = 1$, then a and L can be described in ratios to b.

1.2 The Geometric Definition of Φ

$$a^2 = Lb \quad \text{where} \quad L = a + b$$

Proposition 11 in Book II of *The Elements* is:

"To cut a given straight line so that the rectangle contained by the whole and one of the segments is equal to the square on the remaining segment."

Let us define L as "a given straight line," Lb as "the rectangle contained by the whole and one of its segments," and a^2 as "the square on the remaining segment." In mathematical form, the proposition becomes:

"To cut L so that Lb is equal to a^2."

This is illustrated in the following figure.

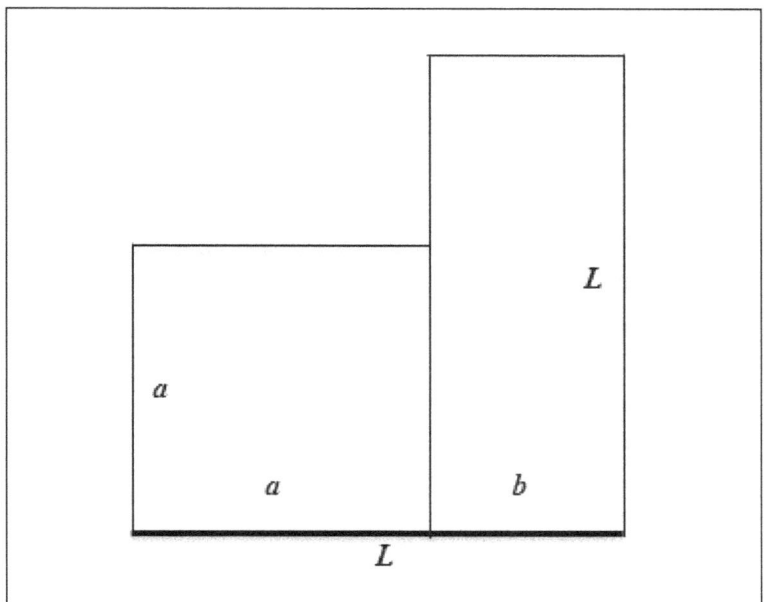

Figure 4: The Geometric Definition of Φ, $a^2 = Lb$

If: $a^2 = Lb$ Equation Area

Then: $\dfrac{L}{a} = \dfrac{a}{b}$

Since $L = a + b$: $\dfrac{L}{a} = \dfrac{a}{b} = \Phi$

Equation Area is sometimes written as $a^2 = L(L - a)$

Notes: In this proposition, Euclid cuts the line in the extreme and mean ratio, without actually calling it that name.

1.3 How to Cut a Line in Φ Proportion

Proposition 30 in Book VI of *The Elements* is:

"To cut a given finite straight line in extreme and mean ratio."

The steps to do this are shown in the following figure.

Step 1: From point 0, on a line perpendicular to line L, go down a distance L/2 to point 1.

Step 2: From point 1, mark the distance h to point 2.

Step 3: Using h as the radius, draw a circular path to point 3.

Step 4: Label the distance from point 0 to point 3 as "*a*." Using "*a*" as the radius, draw a circular path from point 3 to point 4.

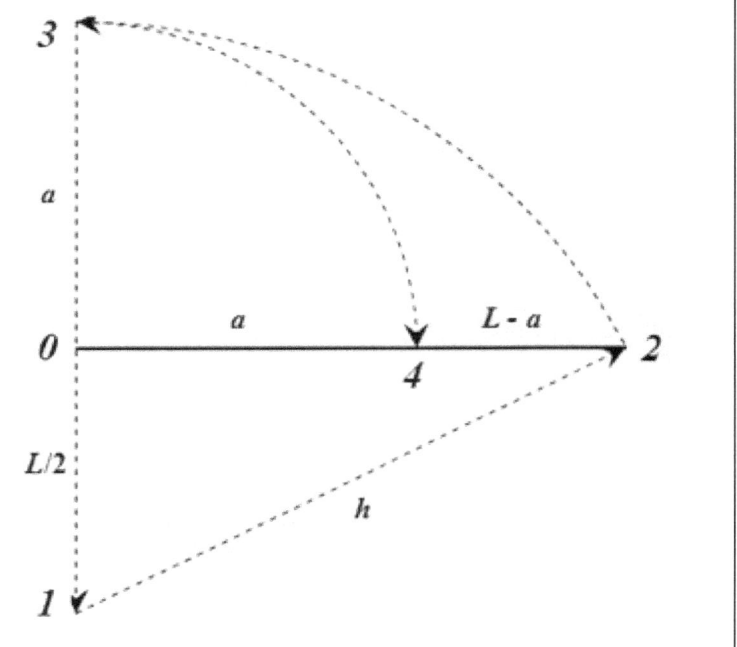

Figure 5: The Φ Cut

The following discussion proves that **L** has been cut at **a** in Φ proportion. Since **h** is the hypotenuse of a right triangle, the Pythagorean theorem* can be used. From the vertical axis in figure 5, $h = (a + L/2)$. Therefore:

$$h^2 = (a + L/2)^2 = (L/2)^2 + L^2$$

This reduces to:

$$(a + L/2)^2 = 5*(L/2)^2$$

Continuing:

$$a^2 + aL + (L/2)^2 = 5*(L/2)^2$$

$$a^2 + aL = L^2$$

Finally:

$$a^2 = L(L - a)$$

This is Equation Area. Hence, **L** has been cut at **a** in Φ proportion.

---

*Euclid proves the Pythagorean theorem in proposition 47 in Book I of *The Elements*. The proposition is: "In right-angled triangles, the square on the side opposite the right angle, equals the sum of the squares on the other two sides." See the appendix for a proof of this theorem.

1.4 The 36/72/72 Triangle and the Regular Decagon

Euclid doesn't specifically construct the regular decagon. It's a by-product of the construction of the 36/72/72 triangle.

Proposition 10 in Book IV of *The Elements* is:

"To construct an isosceles triangle having each of the angles at the base, double of the remaining one."

The method of construction is shown in the following figure.

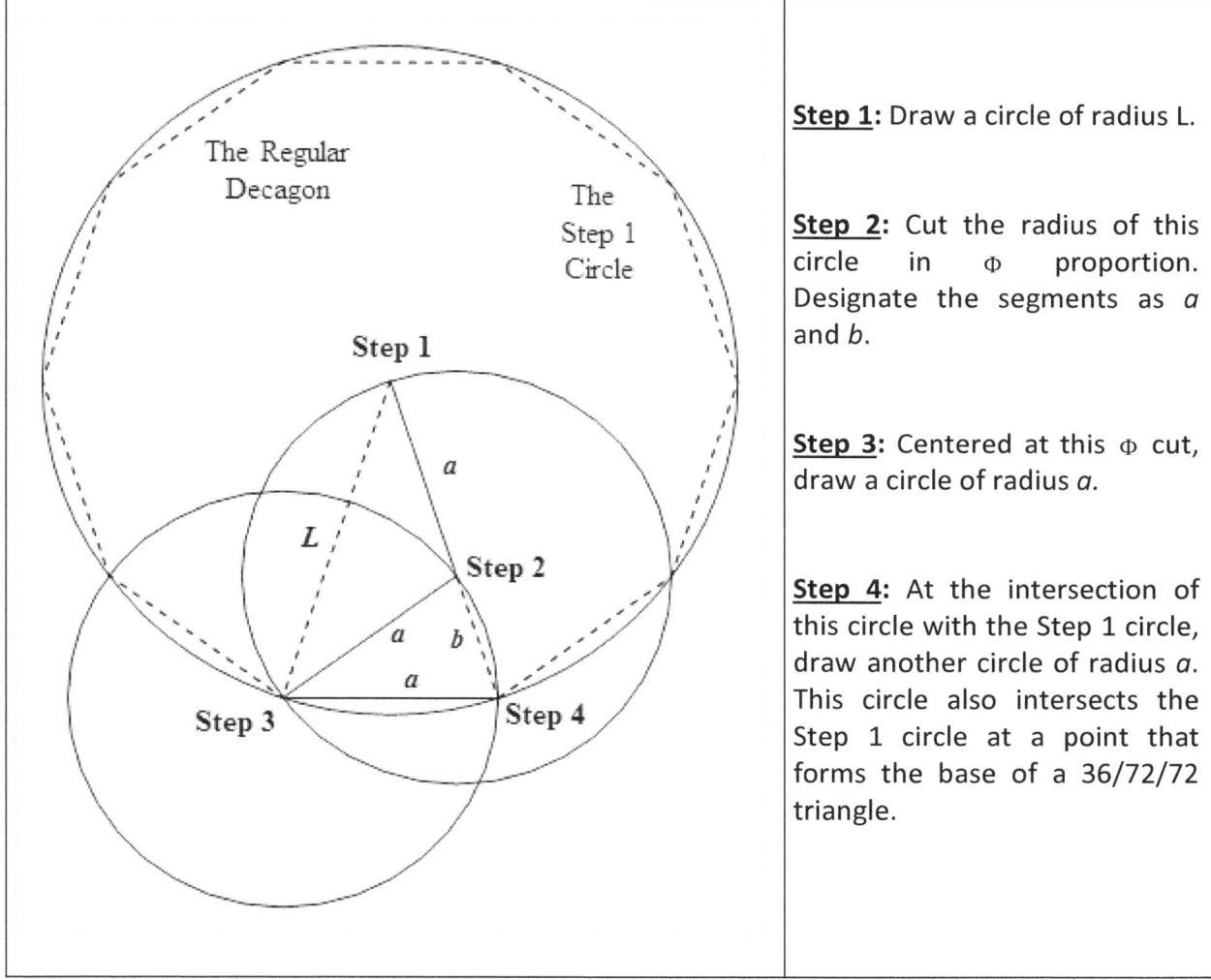

Step 1: Draw a circle of radius L.

Step 2: Cut the radius of this circle in Φ proportion. Designate the segments as *a* and *b*.

Step 3: Centered at this Φ cut, draw a circle of radius *a*.

Step 4: At the intersection of this circle with the Step 1 circle, draw another circle of radius *a*. This circle also intersects the Step 1 circle at a point that forms the base of a 36/72/72 triangle.

Figure 6.a: The 36/72/72 Triangle and the Regular Decagon

Notes: The base of the 36/72/72 triangle is *a*, the greater segment of a line cut in Φ proportion. Each side of the decagon is also "the greater segment." Recall that "the greater segment" is an irrational number.

We will now calculate the angles of the isosceles triangle. Euclid's construction has actually drawn three isosceles triangles, two of them inside the main one. Figure 6.b below, is the main one. Its base angles are called α, and the remaining angle is called β.

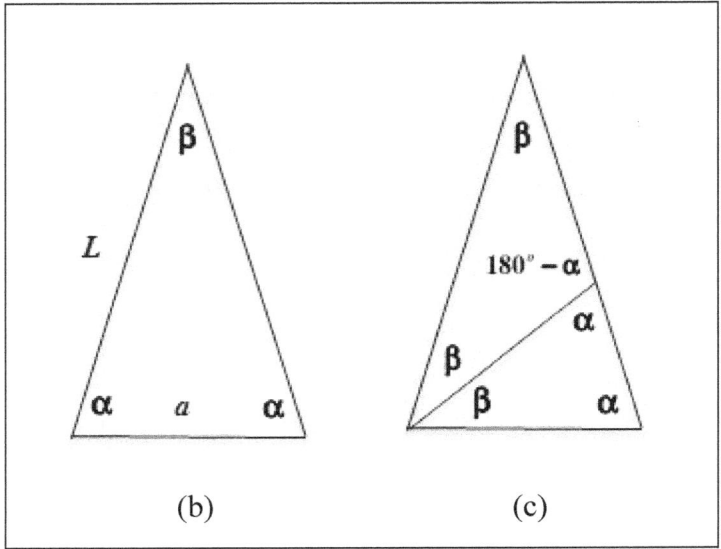

(b)   (c)

Figure 6.b and 6.c: The 36/72/72 Triangle

Figure 6.c shows the two "inside" triangles, along with their angles. The small "inside" triangle has the same angles as the main one. Therefore, the side-to-base ratio of the main one (L/a) is the same as the side-to-base ratio of the small one (a/b).

$$\frac{L}{a} = \frac{a}{b}$$

Since $L = a + b$, that means that

$$\frac{L}{a} = \frac{a}{b} = \Phi$$

In this proposition, Euclid proved geometrically that each of the angles at the base is double the remaining. But he couldn't compute the values of these angles because degree measurements were not available. Using modern math, the degrees of each angle can be easily computed.

Summing the internal angles of the top triangle of figure 6.c yields:

$$\beta + \beta + (180^{o} - \alpha) = 180^{o}$$

From which:
$$\beta = \alpha / 2 .$$

Summing the internal angles of the small triangle: $\alpha + \alpha + \beta = 180^{o}$.

Substituting $\beta = \alpha / 2$, yields $\alpha = 72^{o}$ and $\beta = 36^{o}$. This verifies Euclid's method.

Notes: An isosceles triangle with two of its angles being double the remaining one is a triangle with a side-to-base ratio equal to Φ.

The regular decagon is contained by ten equal lines inscribed in a circle. The ratio of this circle to each of the lines (or sides) is equal to Φ.

1.5 The Regular Pentagon

Proposition 11 in Book IV of *The Elements* is:

"In a given circle, to inscribe an equilateral and equiangular pentagon." The following figure shows a method to construct this pentagon.

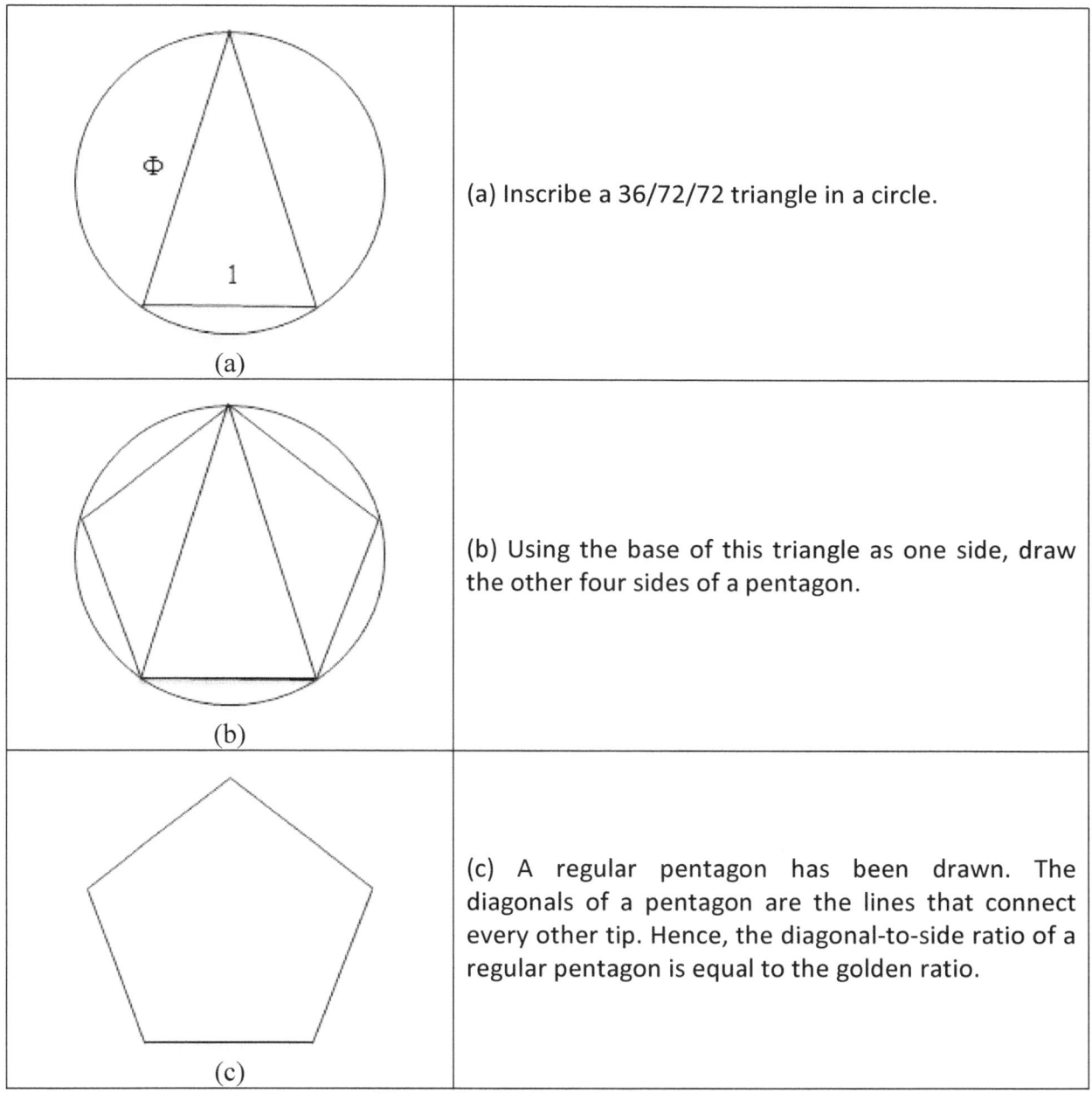

(a)	(a) Inscribe a 36/72/72 triangle in a circle.
(b)	(b) Using the base of this triangle as one side, draw the other four sides of a pentagon.
(c)	(c) A regular pentagon has been drawn. The diagonals of a pentagon are the lines that connect every other tip. Hence, the diagonal-to-side ratio of a regular pentagon is equal to the golden ratio.

Figure 7: The Regular Pentagon

Each side of the pentagon is "the greater segment" and hence is an irrational number.

Notes: A figure is called regular if it is equilateral and equiangular. As such, it can be inscribed in a circle or sphere.

1.6 The Platonic Solids

Book XIII of *The Elements* is devoted to the construction of the five figures known as the Platonic solids. These figures are shown below.

Pythagoras (c.a. 500 BC) knew of at least three of these figures. But in 390 BC, Plato wrote about them in his book *Timaeus*. That's why they are called the Platonic solids. In *Timaeus*, Plato says that these solids represent earth, air, fire, and water and the shape of the universe. This has always linked these solids to the Divinity.

The Platonic solids are categorized as regular polyhedrons. They are *regular* because their identical faces are equilateral and equiangular. They are *polyhedrons* because they are three-dimensional figures that are formed by planar faces. In proposition 18 of Book XIII, Euclid proves that there can only be five kinds of Platonic solids.

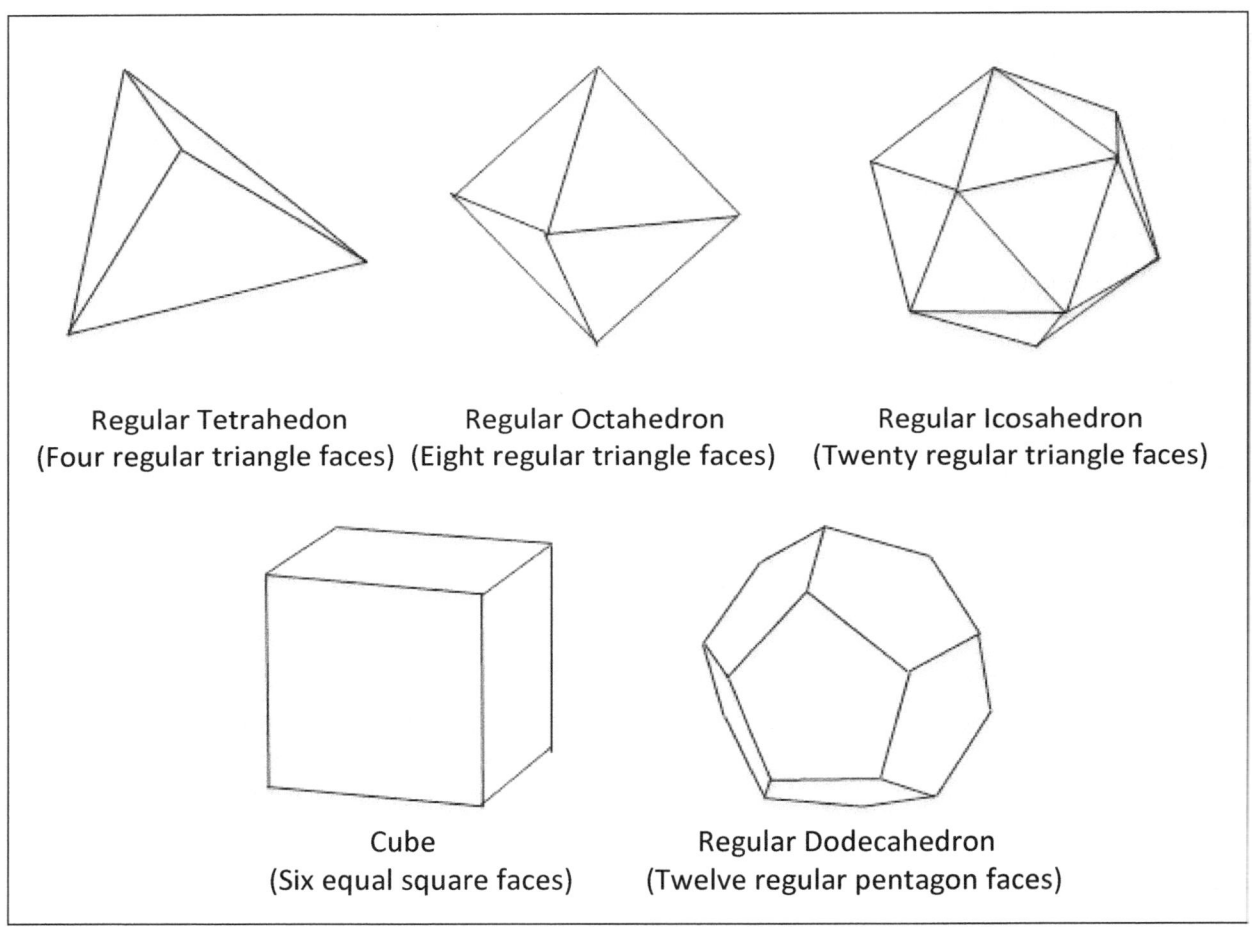

Regular Tetrahedon   Regular Octahedron   Regular Icosahedron
(Four regular triangle faces) (Eight regular triangle faces) (Twenty regular triangle faces)

Cube   Regular Dodecahedron
(Six equal square faces)   (Twelve regular pentagon faces)

Figure 8: The Platonic Solids

As will be shown in the next sections, the golden ratio is literally the heart of the icosahedron and the dodecahedron.

Notes: Euclid referred to the regular tetrahedron as a pyramid.

1.7 The Golden Rectangle

This figure was not constructed by Euclid in a separate proposition. But it has become a part of the modern literature on the golden ratio. And it is a central element of the modern method for constructing the regular icosahedron and the regular dodecahedron.

The following figure shows the golden rectangle. It's a rectangle constructed on the greater segment, *a*, of a line cut in extreme and mean ratio.

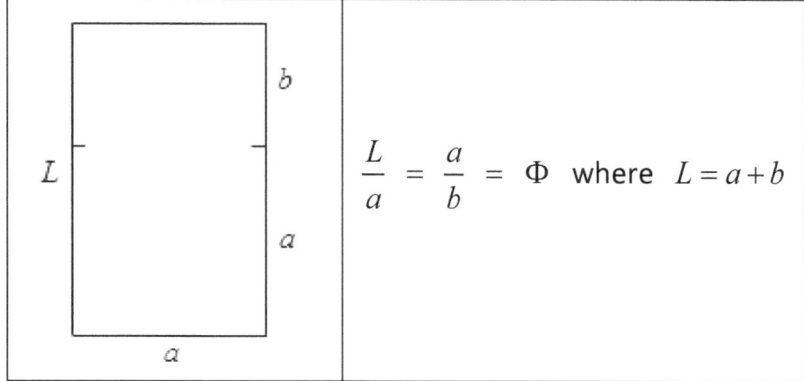

Figure 9.a: The Golden Rectangle

The following figure shows a rectangle constructed on the lesser segment, *b*, of a line cut in extreme and mean ratio. Let's call it the Φ^2 rectangle.

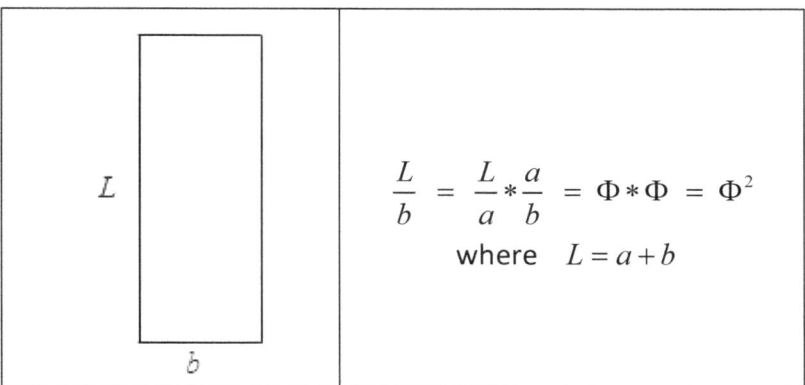

Figure 9.b: The Φ^2 Rectangle

The golden rectangle is used to construct the regular icosahedron. The Φ^2 rectangle is used to construct the regular dodecahedron.

1.8 The Regular Icosahedron

Proposition 16 in Book XIII of *The Elements* begins with:

"To construct an icosahedron and comprehend it in a sphere..."

The following figure shows a modern-day method of construction.

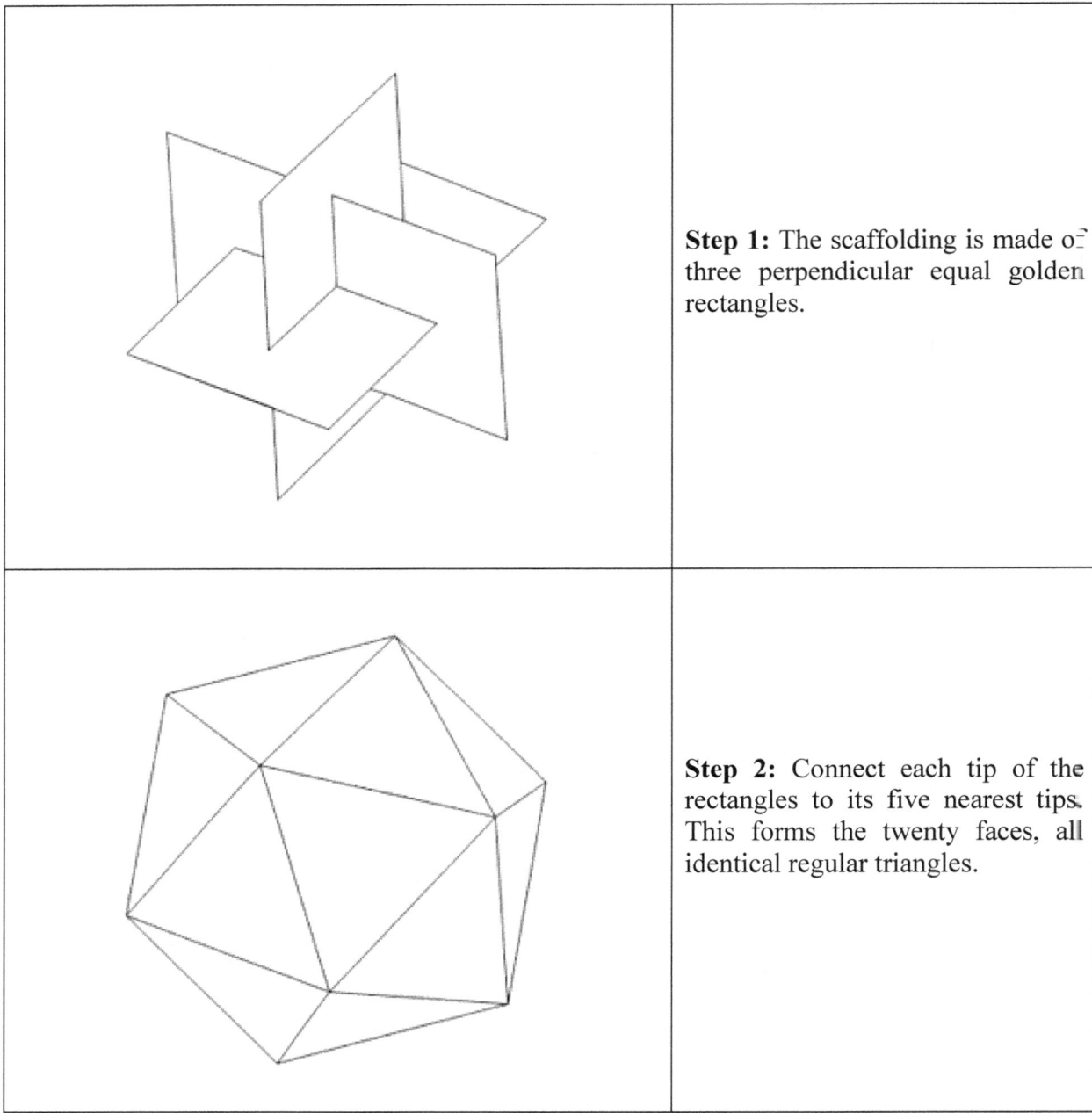

Step 1: The scaffolding is made of three perpendicular equal golden rectangles.

Step 2: Connect each tip of the rectangles to its five nearest tips. This forms the twenty faces, all identical regular triangles.

Figure 10: The Regular Icosahedron

Notes: Each edge of this icosahedron is the greater segment of a line cut in Φ proportion and hence is an irrational number.

1.9 The Regular Dodecahedron

Proposition 17 in Book XIII of *The Elements* begins with:

"To construct a dodecahedron and comprehend it in a sphere…"

The following figure shows a modern-day method of construction.

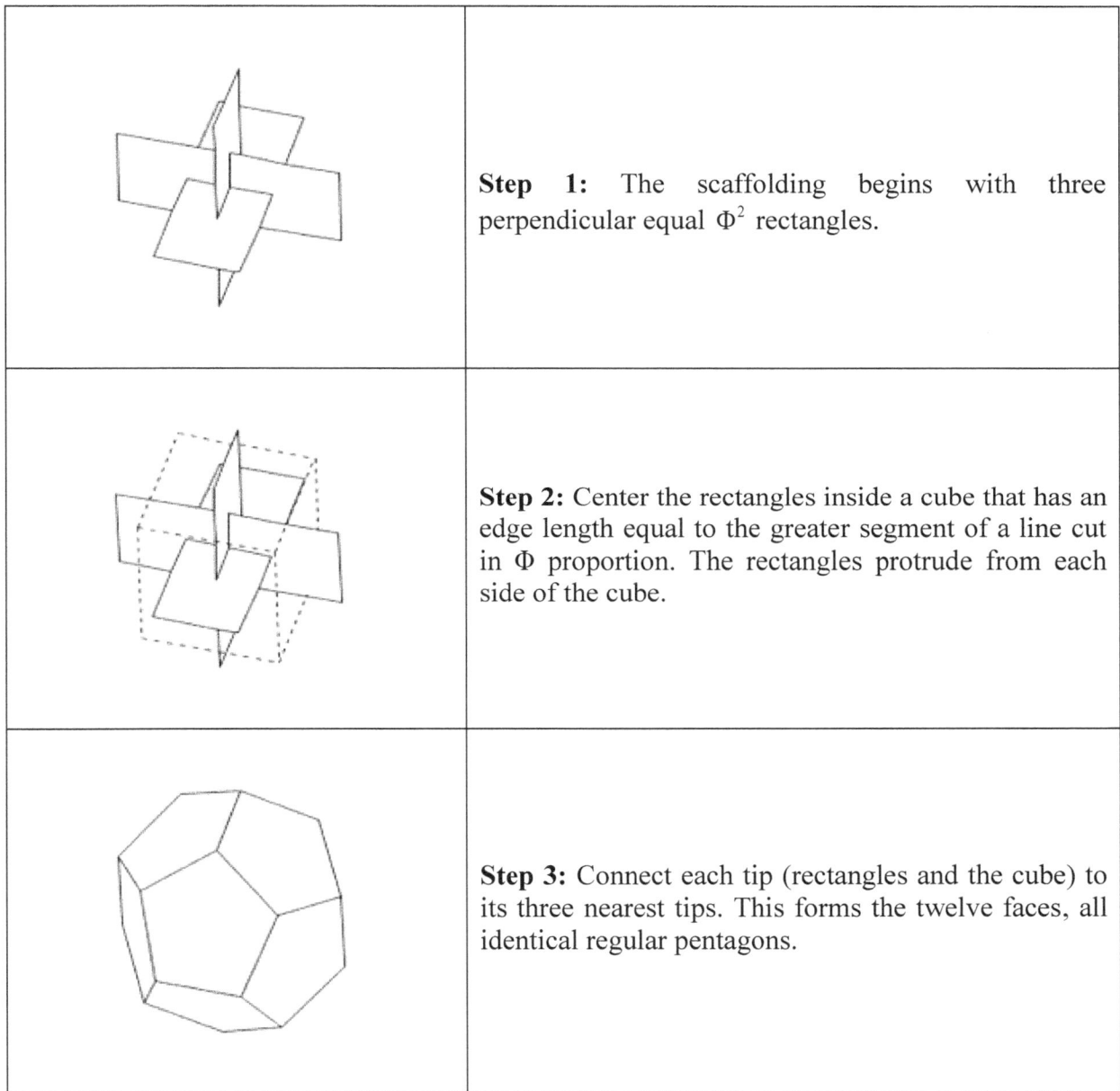

	Step 1: The scaffolding begins with three perpendicular equal Φ^2 rectangles.
	Step 2: Center the rectangles inside a cube that has an edge length equal to the greater segment of a line cut in Φ proportion. The rectangles protrude from each side of the cube.
	Step 3: Connect each tip (rectangles and the cube) to its three nearest tips. This forms the twelve faces, all identical regular pentagons.

Figure 11: The Regular Dodecahedron

Notes: The edge of this dodecahedron is the lesser segment of a line cut in Φ proportion and hence is an irrational number. The diagonal of each pentagon face is the greater segment, the same as the edge of the cube.

1.10 Modern-Day Arithmetic Properties of Φ

The following table shows ways of representing Φ that were not possible in Euclid's time.

- To compute its square, add *unity*. $\qquad \Phi^2 = \Phi + 1$

- To compute its inverse, subtract *unity*. $\qquad \dfrac{1}{\Phi} = \Phi - 1$

- Subtract it from its square and get *unity*. $\qquad \Phi^2 - \Phi = 1$

- Don't forget that Φ is also ... $\Phi = -0.618...$

Chapter 2: The Myths about Φ

A myth is a popular belief about something. Some myths are soundly based. Some are not. Belief is a choice. This chapter analyzes the important myths.

2.1 *The Divine Proportion*: This is the name of a math textbook by Luca Pacioli. Today, we credit Pacioli as being the Renaissance man, as well as being the father of accounting. This legendary monk was a frequent consultant to Leonardo da Vinci on mathematics. In Pacioli's textbook, he lists five reasons why he considered Φ the "divine proportion." One of the reasons is that Φ is an irrational number.

2.2 Kepler's Solar System: Kepler proved that the planets go around the sun in elliptical orbits. But in earlier studies, he theorized that each planet rolled around the sun on invisible concentric spheres. He used the Platonic solids as spacers between these spheres. And two of these solids are constructed using Φ.

2.3 The Fibonacci Numbers: Leonardo of Pisa, also known as Fibonacci, is credited with introducing arabic numbers into Europe. His landmark textbook, *Liber Abaci*, or *The Book of Numbers*, convinced European merchants that arabic numbers were better than roman numbers. In this textbook, he introduced a series of numbers that today are called the Fibonacci numbers. Four hundred years after Leonardo of Pisa, Kepler discovered that these numbers have a link to Φ.

2.4 The Golden Spiral: This is the heart of the myth that claims that Φ is used in many patterns of nature. It seems to validate the belief that Φ is a tool of God. This section shows that the spiral created by arrangements of the golden rectangle resembles the spiral seen in snail shells.

2.5 The Pyramids: The pyramids have always been associated with divinity. And they look like they could have been constructed using Φ. The pyramids predate Euclid by thousands of years, so that would indeed make Φ a mythical tool.

2.6 The Pythagorean Star: Pythagoras is known as the godfather of math. His academy in philosophy and mathematics was called the Pythagorean Brotherhood. Its motto was "Everything is number." Its emblem was what we now call the Pythagorean star. Was this star constructed using Φ?

The chapter headings show that Φ is associated with some important parts of history. No doubt this has magnified the myths about Φ.

2.1 *The Divine Proportion*

In the early days of the Renaissance, there was a very strong connection among religion, science, and mathematics. In 1500, around the start of the Renaissance, Luca Pacioli wrote a masterpiece textbook called *Summa*. It was a collection of the math knowledge of the time. It covered arithmetic, algebra, geometry, and trigonometry. Because of this book, Pacioli is still recognized as the father of accounting.

In 1509, Pacioli wrote another book called *The Divine Proportion*. This was the name he gave to Φ. Pacioli lists five reasons why. They are: (1) the ratio is unique like God, (2) it has three lengths like the Trinity of God, (3) it is incomprehensible like God because its numerical value is an irrational or "unknowable" number, (4) it is invariable like God since its value doesn't depend on the length of the line being divided, and (5) it is related to the Platonic solids that symbolize earth, air, fire, water, and the shape of the universe. (This list of reasons is paraphrased from Mario Livio.[4])

Pacioli was also a high-ranking friar and had papal privileges. This is how he came to be a math consultant to Leonardo da Vinci. He also got da Vinci to illustrate *The Divine Proportion*. Da Vinci provided sixty illustrations, including the ones shown in figure 12.

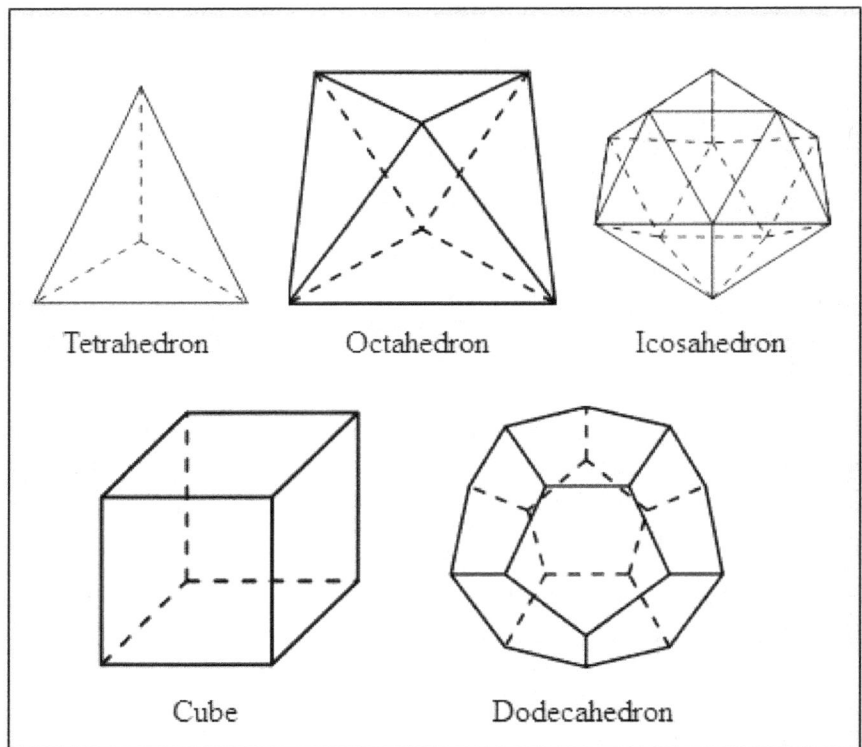

Figure 12: da Vinci's Platonic Solids in *The Divine Proportion*

---

4. Mario Livio, *The Golden Ratio* (New York: Broadway Books, 2002).

2.2 Kepler's Solar System

In 1543, Copernicus shocked the world with his claim that the planets revolved around the sun in circular orbits. Then, in 1609, Kepler published a book with data showing that the orbits were elliptical.

Earlier, in 1597, Kepler had published a book titled *Mysterium Cosmographicum*. He was trying to figure out why the orbits were circular, as Copernicus had claimed. Here, he made his now famous statement that "God ever geometrizes." Not knowing about gravity, Kepler theorized that the planets rolled around the sun on invisible spheres. He said that invisible spacers between the spheres kept everything together.

It is known that Kepler studied *The Elements*. It is likely he got his *spacer* idea from proposition 17 of Book XII, which says: "Given two spheres about the same centre, to inscribe in the greater sphere a polyhedral solid which does not touch the lesser sphere at its surface."

Kepler's idea was that God had used a model like this to build the solar system. At the time, there were six known planets. The five Platonic solids are polyhedrons. The model of Kepler's idea is shown in figure 13, which clearly shows the cube as the spacer between the orbits of Saturn and Jupiter. The rest of the model is described in figure 13.

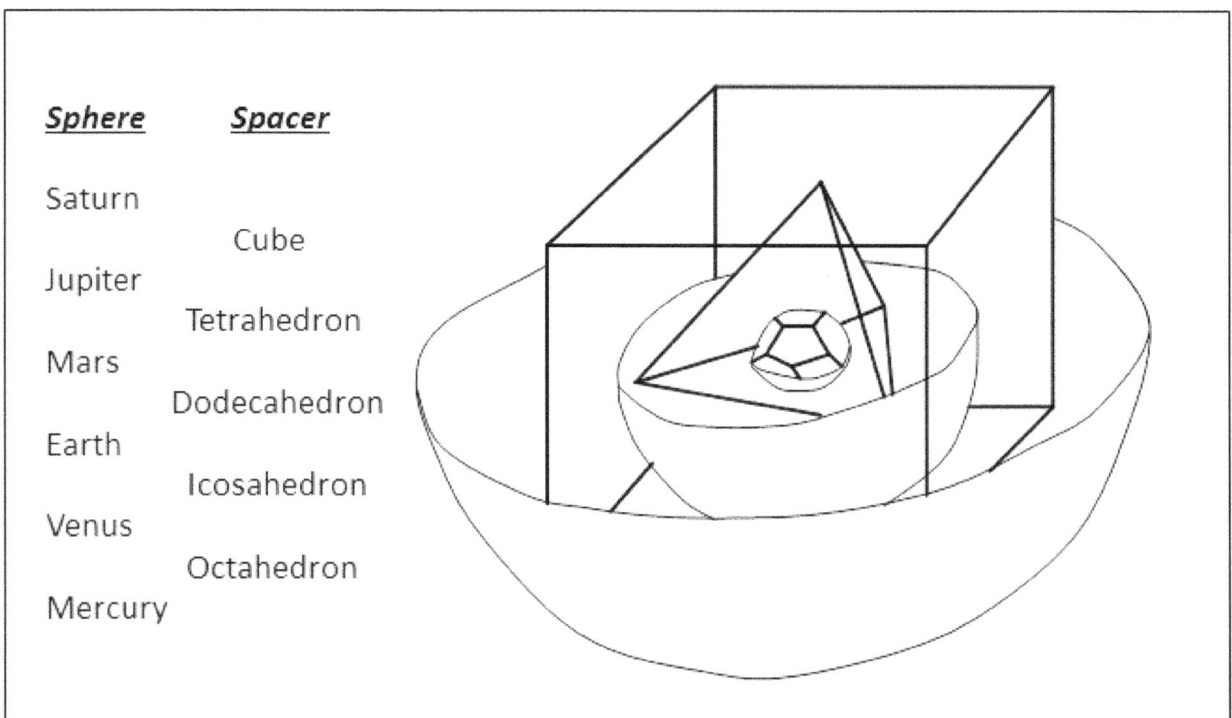

Figure 13: Kepler's Solar System in *Mysterium Cosmographicum*

Nobody, including Kepler, really believed in this crazy contraption. Kepler's further studies soon showed that the planets move in elliptical orbits. In 1609, he wrote *New Astronomy*, revealing his now famous, and still valid, Kepler's laws of orbital motion. It has to be said again: Kepler did this without knowing about *gravity*. That discovery would come eighty years later from Sir Isaac Newton.

2.3 The Fibonacci Numbers

These numbers are named after a man whose story is one of the most remarkable in Western history.

Around 1200, Europe was still in the Dark Ages. Leonardo of Pisa, also known as Fibonacci, was the son of an Italian businessman. As a boy, he went on his father's business trips throughout the Arab countries. He observed that the Arabs had a huge negotiating advantage over his father, because they were using the arabic number system, while his father was still using roman numerals.

When Fibonacci grew up and began his academic career, he started to teach the arabic number system to his students in Italy. He wrote a textbook on business math using these numbers. It was called *Liber Abaci*, or *The Book of Numbers*. This textbook was an enormous success and soon made Fibonacci one of Europe's preeminent mathematicians. Fibonacci is credited with introducing arabic numbers into Europe.

As a textbook, *Liber Abaci* had example problems. One of the problems was to predict how rabbits propagate. Fibonacci said that rabbits propagate according to the numbers that are given by his now famous Fibonacci equation: $F_n = F_{n-1} + F_{n-2}$. That is, the current number in the series is the sum of the last two numbers. The first ten numbers in this series are plotted in figure 14.

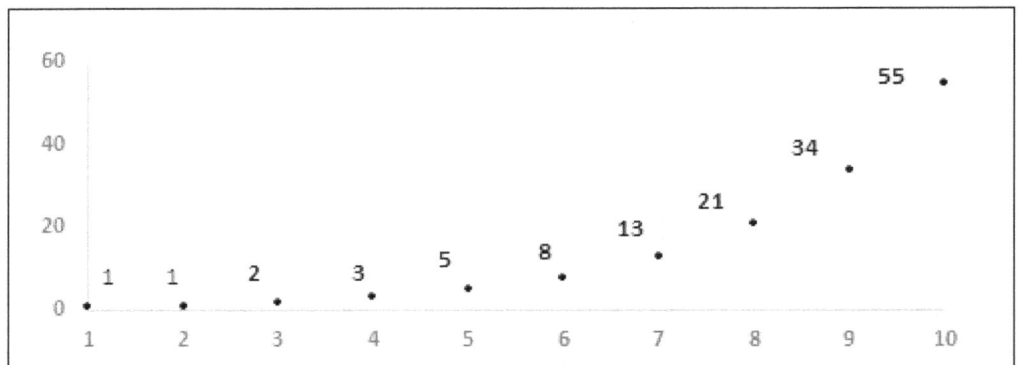

Figure 14: The First Ten Fibonacci Numbers

To Fibonacci, this was nothing more than an example problem in his textbook. In fact, because it was a twelve-month rabbit-propagation problem, only the first twelve numbers appear in *Liber Abaci*.

Around 1600, Kepler (of solar system fame) was studying Euclid's *The Elements*, Pacioli's *The Divine Proportion*, and Fibonacci's *Liber Abaci*. For his reasons, Kepler computed the ratio of successive Fibonacci numbers. These are plotted in figure 15.

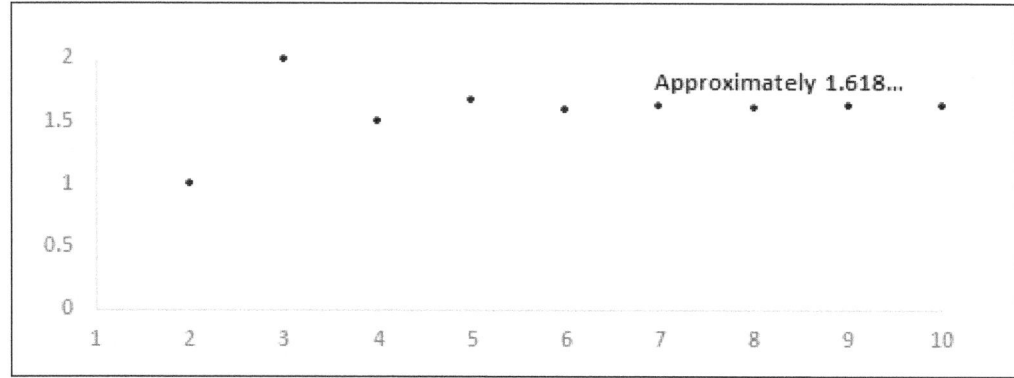

Figure 15: Ratio of Successive Fibonacci Numbers

Kepler saw that as the numbers get larger, this ratio approaches the value of 1.618. This is the value of Φ. And Kepler regarded Φ as the "jewel of geometry." Again Φ is associated with a famous person, and the myth grows. But let's study this a little bit.

Around 1800, Edouard Lucas, a prominent French mathematician, published another famous series of numbers. This bears his name. His series uses the Fibonacci equation, but with different starting values ($F_1 = 1$ and $F_2 = 3$). This series is plotted in the following figure.

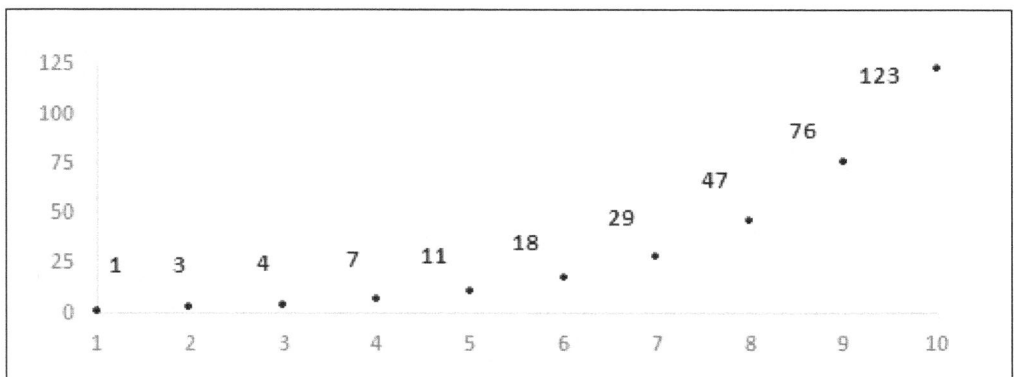

Figure 16: The First Ten Lucas Numbers

Figure 17 is a plot of the ratio of successive Lucas numbers. This ratio also converges to 1.618. This shows that convergence to 1.618 depends on the equation, and not on the numbers themselves.

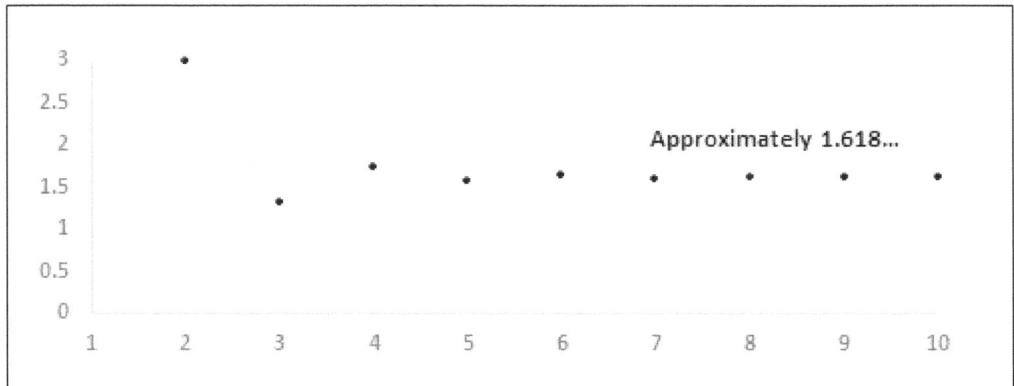

Figure 17: Ratio of Successive Lucas Numbers

The Fibonacci and Lucas number sequences are generated by the equation:

$$F_n = F_{n-1} + F_{n-2}$$

As with the line that Euclid used to define Φ, the whole number F_n is the sum of two segments, F_{n-1} and F_{n-2}. The ratio of the whole number to the large segment is F_n / F_{n-1}. The ratio of the large segment to the small segment is F_{n-1} / F_{n-2}. Figures 15 and 17 show that for large numbers, these ratios approach the same value. The following discusses why this happens.

The formula for the ratio of successive numbers is:

$$\frac{F_n}{F_{n-1}} = 1 + \frac{F_{n-2}}{F_{n-1}}$$

Manipulating yields:
$$\frac{F_n}{F_{n-1}} = 1 + \frac{1}{(F_{n-1} / F_{n-2})} \qquad \text{Equation F}$$

As the numbers get large:
$$\frac{F_n}{F_{n-1}} \approx \frac{F_{n-1}}{F_{n-2}}$$

For large numbers, let:
$$R = \frac{F_n}{F_{n-1}} = \frac{F_{n-1}}{F_{n-2}}$$

Substituting R into equation F:
$$R = 1 + \frac{1}{R} \qquad \text{to yield } R^2 - R - 1 = 0$$

The solution of this is R = 1.618, which happens to equal the value of Φ. That's the math that explains the ratio of successive Fibonacci numbers.

It is not known if Fibonacci knew that his number sequence is related to Φ. What is known is that Fibonacci knew about the extreme and mean ratio. He wrote another book, titled *The Practice of Geometry*. In that book, Fibonacci used the extreme and mean ratio to construct the same figures as Euclid.

2.4 The Golden Spiral

Quoting from Mario Livio: "Kepler truly believed that the Golden Ratio served as a fundamental tool for God in creating the universe."[5] An example pertinent to this is the geometry of a seashell or snail's shell.

The golden rectangle is defined in chapter 1.7. In the figure below, the top three figures are golden rectangles, arranged to show how the side of one can lead to the side of another. In the bottom of this figure, these rectangles are fit together.

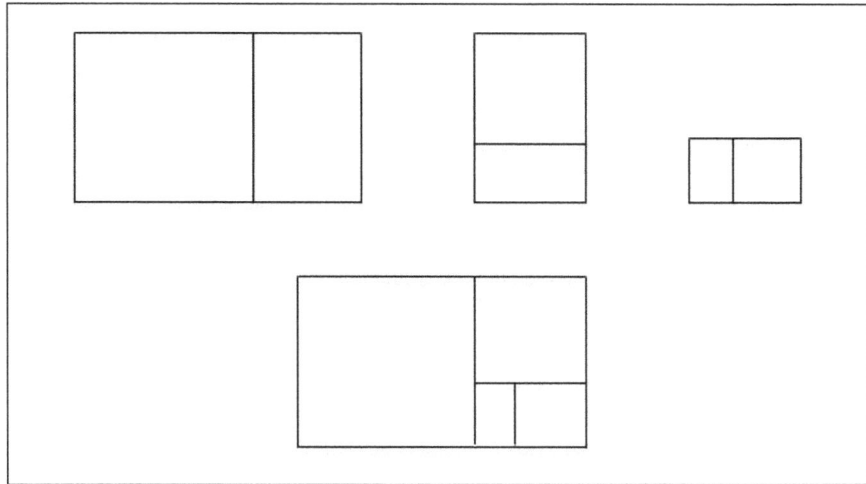

Figure 18: Golden Rectangle Patterns

Now look at the three figures at the top of figure 19. They are the same as figure 18, with a quarter circle drawn in the square areas of each. The bottom of figure 19 shows the rectangles fit together, forming a spiral. It does resemble a seashell or a snail. By adding more rectangles, the spiral can continue indefinitely.

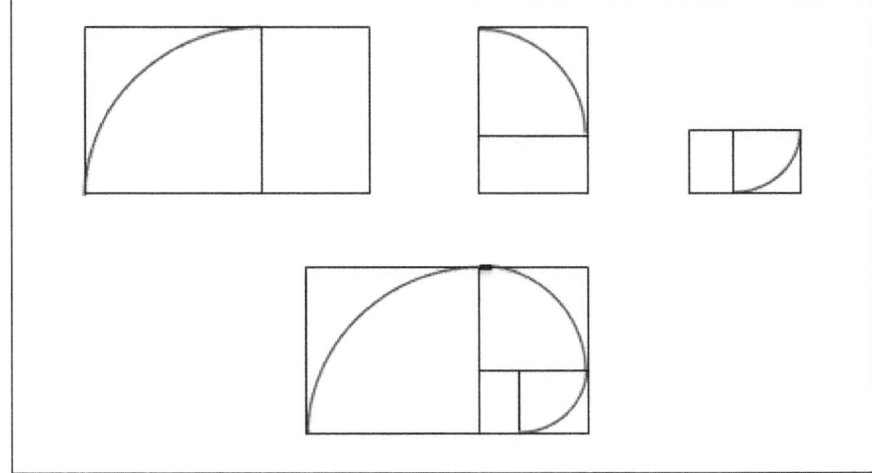

Figure 19: The Golden Spiral

---

5. Mario Livio, *The Golden Ratio* (New York: Broadway Books, 2002).

The golden spiral is often cited as an example of the correlation between the laws of mathematics and the structure of living things.

The golden spiral is also often called the logarithmic spiral. This is not mathematically correct. Figure 20 compares the golden spiral (dashed curve) to the logarithmic spiral (solid curve).

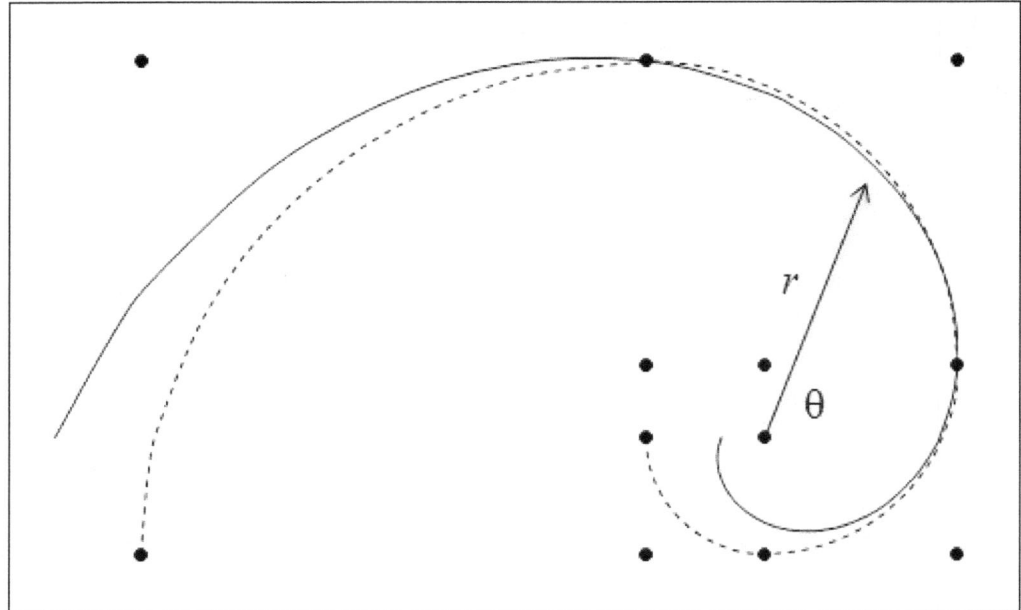

Figure 20: Comparing the Golden Spiral to the Logarithmic Spiral

The magnitude (r) of the logarithmic spiral is computed from the following equation:

$$r = A * e^{B*\theta}$$

In this formula, θ is the angle in radians, e is the exponential function, and A and B are selectable constants. For figure 20, A and B were chosen to match the golden spiral that is shown. Because the logarithmic spiral has only two selectable constants, it can exactly match the golden spiral at only two values. Therefore the match can only be good in a region.

Notes: In this generic case, the values of A and B are not important. However, they are given for completeness: $A = 0.56$ and $B = 0.45$.

2.5 The Pyramids

The pyramids have always been associated with divinity. They look like they could have been constructed using the golden ratio. That would indeed be mythical, since they were built two thousand years before Φ was documented by Euclid. The following figure depicts a facial view of the Khufu pyramid, the world's largest. The ratio of its edge (L) to its base (a) is:

$$\frac{L}{a} = 0.81$$

	The Khufu Pyramid (Facial View) 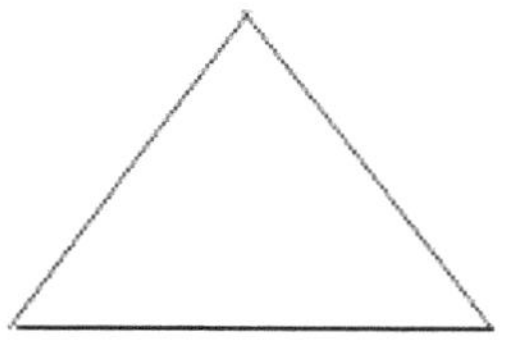	A *Golden Pyramid* (Facial View) 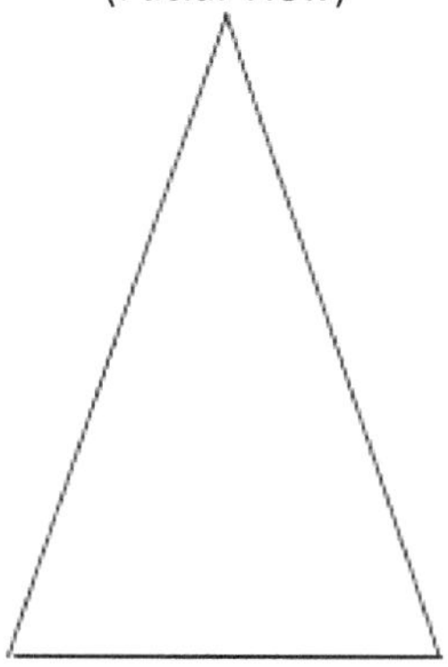
Edge-to-base ratio	0.81	1.618
Base (feet)	755	755
Edge (feet)	612	1222
Height (feet)	480	1106
Base angle (degrees)	52	72

Figure 21: The Facial View of the Pyramids

Figure 21 also shows a pyramid with an edge-to-base ratio of Φ. No matter how you look at the Khufu pyramid, it was not likely constructed using Φ. This conclusion is the same for the other known pyramids.

Notes: In proposition 13 of Book XIII, Euclid refers to the tetrahedron as a pyramid. Φ is not used in the construction of the tetrahedron.

2.6 The Pythagorean Star

Pythagoras lived about two hundred years before Euclid. He was one of the first great mathematicians in the Western world. He established a secret academy in philosophy and mathematics called the Pythagorean Brotherhood. This academy made many remarkable discoveries in math, but its famous Pythagorean theorem was known by the Babylonians much earlier. Euclid used this theorem often. The appendix contains a proof of this theorem.

Around AD 450, Proclus, the preeminent Greek historian, credited Pythagoras with transforming math from practical use to a theoretical liberal art. The motto of the Pythagorean Brotherhood was "Everything is number." The Pythagorean star was the emblem of the Brotherhood. Figure 22 shows one way to construct this star.

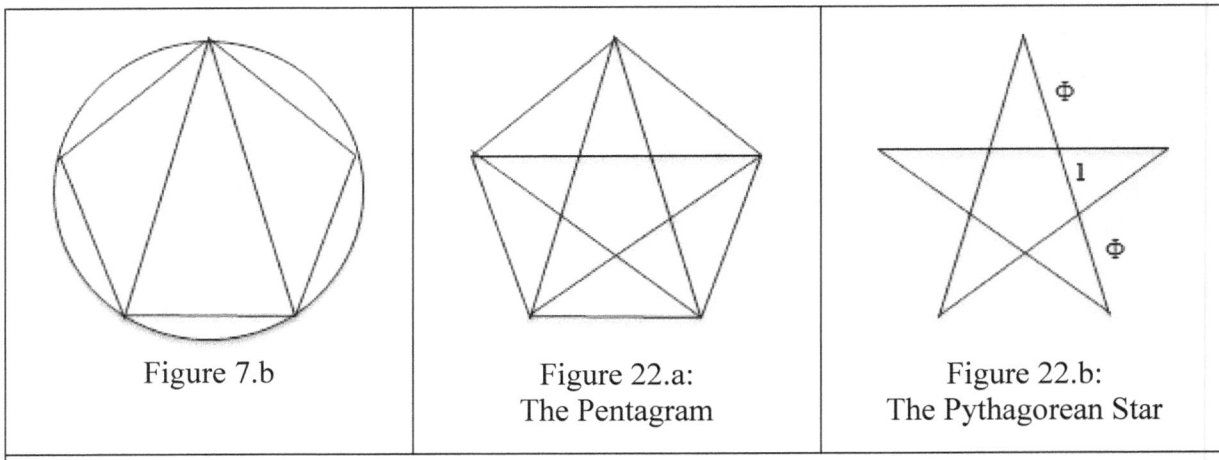

| Figure 7.b | Figure 22.a: The Pentagram | Figure 22.b: The Pythagorean Star |

- Figure 7.b is repeated from chapter 1.5 on the pentagon.

- Figure 22.a is figure 7.b with the circle removed and all diagonals added. This figure is the pentagram.

- Figure 22.b is the pentagram with the pentagon removed, and is the Pythagorean star.

Figure 22: The Pythagorean Star, the Emblem of the Pythagorean Brotherhood

Figure 22 shows a connection between Pythagoras and Φ. The Pythagorean star is frequently drawn showing Φ as the length of its tips. The tips are connected by diagonals of a regular pentagon. The following is one of the ways to represent these diagonals (see figure 3).

$$L = \Phi + 1 + \Phi$$

But there's no proof that Pythagoras knew about this connection with Φ. Also, Euclid did not construct either the pentagram or the Pythagorean star.

Chapter 3: Book XIII and Φ

All of Book XIII deals with the Platonic solids. Recall that these are regular polyhedrons, which are 3-D figures contained by equal straight lines that form equal faces. Thirteen of the eighteen propositions in Book XIII involve Φ. This chapter discusses these thirteen.

Euclid constructs each Platonic solid using scaffolding that consists of straight lines. The lines are cut to various lengths using relations among polygons and circles that are derived in many propositions. As an example, Euclid frequently uses the Pythagorean theorem (proposition 47 in Book I), which relates squares that are constructed on the sides of a right triangle.

Propositions 1 to 5 relate squares and rectangles that are constructed on lines cut in Φ proportion. These propositions are good examples of how Euclid did algebra using geometry.

Propositions 6 to 12 derive other relationships that Euclid used to construct the Platonic solids. These relationships show how Euclid handled irrational numbers.

Propositions 13, 14, and 15 show the construction of the regular tetrahedron (pyramid), octahedron, and cube. These do not involve Φ.

Proposition 16 and 17 show the construction of the regular icosahedron and dodecahedron.

Proposition 18 is the famous proposition wherein Euclid proves that the Platonic solids are the only regular polyhedrons that can exist.

3.1 Proposition 1

From *The Elements*: "If a straight line be cut in extreme and mean ratio, the square on the greater segment added to the half of the whole is five times the square on the half."

We will begin by defining:

1. "A straight line" as $L = a + b$, with the greater segment being a.
2. "The greater segment added to the half of the whole" as $(a + L/2)$.
3. "The half" as $L/2$.

Using mathematical notation, this proposition is:

"If L be cut in extreme and mean ratio, $(a + L/2)^2 = 5 * (L/2)^2$."

Note that this equation was encountered in section 1.3 (proposition 30 in Book VI), which dealt with proportioning a line in the extreme and mean ratio.

The following table shows the geometric solution:

Step 1: Erect a square on the line segment (a + L/2) and form four areas.

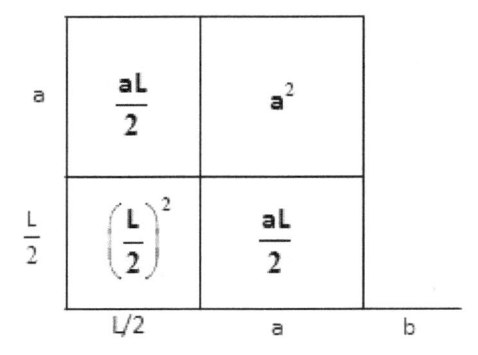

- $\left(a + \dfrac{L}{2}\right)^2 = a^2 + \dfrac{aL}{2} + \dfrac{aL}{2} + \left(\dfrac{L}{2}\right)^2$

Step 2: Erect a square on the line (L = a + b) and form two areas.

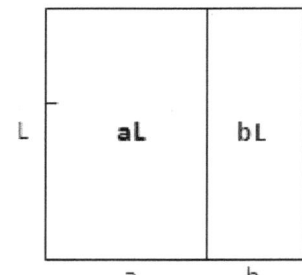

- $L^2 = bL + aL$
- From proposition II-11, $bL = a^2$
- Therefore: $L^2 = a^2 + aL$

Step 3: Substitute L^2 into $\left(a + \dfrac{L}{2}\right)^2$: $\quad \left(a + \dfrac{L}{2}\right)^2 = L^2 + \left(\dfrac{L}{2}\right)^2 = 5*\left(\dfrac{L}{2}\right)^2$.

Notes: Euclid concluded most of his propositions with Q.E.D. This means "*quod erat demonstrandum,*" or "*which was to be demonstrated.*"

The following table shows the algebraic solution:

- $\left(a + \dfrac{L}{2}\right)^2 = a^2 + aL + \left(\dfrac{L}{2}\right)^2$
- From proposition II-11, $a^2 = bL$
- Therefore: $\left(a + \dfrac{L}{2}\right)^2 = bL + aL + \left(\dfrac{L}{2}\right)^2 = L^2 + \left(\dfrac{L}{2}\right)^2 = 5*\left(\dfrac{L}{2}\right)^2$

3.2 Proposition 2

From *The Elements*: "If the square on a straight line be five times the square on a segment of it, then, when the double of the said segment is cut in extreme and mean ratio, the greater segment is the remaining part of the original straight line."

We will begin by defining:

1. "A straight line" as $(a + L/2)$.
2. "A segment of it" as $L/2$.
3. "The double of the said segment" as L.
4. "The greater segment" as a.
5. "The original straight line" as $(a + L/2)$.

Using mathematical notation, this proposition is:

"If $(a + L/2)^2 = 5 * (L/2)^2$, then, when L is cut in extreme and mean ratio, a is the remaining part of $(a + L/2)$."

This is the converse of proposition 1.

3.3 Proposition 3

From *The Elements*: "If a straight line be cut in extreme and mean ratio, the square on the lesser segment added to the half of the greater segment is five times the square on the half of the greater segment."

We will begin by defining:

1. "A straight line" as L, with a greater segment **a**, and a lesser segment **b**.
2. "The lesser segment added to the half of the greater segment" as $(b + a/2)$.
3. "The half of the greater segment" as $a/2$.

Using mathematical notation, the proposition is:

"If L be cut in extreme and mean ratio, $(b + a/2)^2 = 5*(a/2)^2$."

The following table shows the geometric solution.

Erect a square on the line segment (b + a/2) and form four areas.	
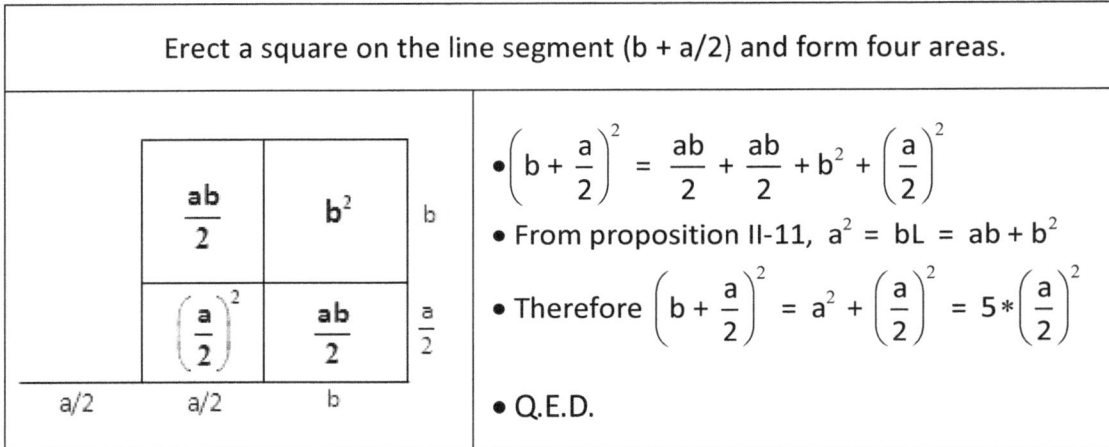	• $\left(b + \dfrac{a}{2}\right)^2 = \dfrac{ab}{2} + \dfrac{ab}{2} + b^2 + \left(\dfrac{a}{2}\right)^2$ • From proposition II-11, $a^2 = bL = ab + b^2$ • Therefore $\left(b + \dfrac{a}{2}\right)^2 = a^2 + \left(\dfrac{a}{2}\right)^2 = 5*\left(\dfrac{a}{2}\right)^2$ • Q.E.D.

The following table shows the algebraic solution.

• $\left(b + \dfrac{a}{2}\right)^2 = b^2 + ab + \left(\dfrac{a}{2}\right)^2 = b(b + a) + \left(\dfrac{a}{2}\right)^2 = bL + \left(\dfrac{a}{2}\right)^2$ • From proposition II-11, $bL = a^2$ • Therefore: $\left(b + \dfrac{a}{2}\right)^2 = a^2 + \left(\dfrac{a}{2}\right)^2 = 5*\left(\dfrac{a}{2}\right)^2$

3.4 Proposition 4

From *The Elements*: "If a straight line be cut in extreme and mean ratio, the square on the whole and the square on the lesser segment together are triple of the square on the greater segment."

We will begin by defining:

1. "A straight line" as L.
2. "The whole" as L.
3. "The lesser segment" as **b**.
4. "The greater segment" as **a**.

Using mathematical notation, this proposition is:

"If L be cut in extreme and mean ratio, $L^2 + b^2 = 3*a^2$."

The following table shows the geometric solution.

Erect a square on the line segment $(a + b)$, another one on the line segment (b), and form five areas.	
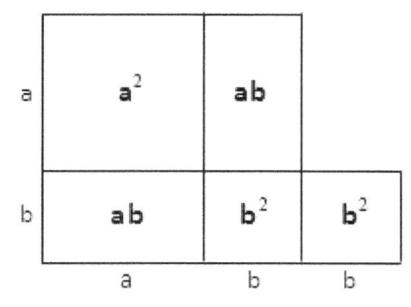	• $L^2 + b^2 = ab + ab + b^2 + b^2 + a^2$ $\quad = 2(ab + b^2) + a^2$ • From proposition II-11, $a^2 = bL = ab + b^2$ • Therefore: $L^2 + b^2 = 2a^2 + a^2 = 3a^2$ • Q.E.D.

The following table shows the algebraic solution.

• $L^2 + b^2 = (a + b)^2 + b^2 = a^2 + 2ab + b^2 + b^2 = a^2 + 2(ab + b^2)$ • From proposition II-11, $a^2 = bL = ab + b^2$ • Therefore: $L^2 + b^2 = a^2 + 2a^2 = 3a^2$

3.5 Proposition 5

From *The Elements*: "If a straight line be cut in extreme and mean ratio, and there be added to it a straight line equal to the greater segment, the whole straight line has been cut in extreme and mean ratio, and the original straight line is the greater segment."

We will begin by defining:

1. "A straight line" as L.
2. "A straight line equal to the greater segment" as **a**.
3. "The whole straight line" as $L + a$.
4. "The original straight line" as L.

Using mathematical notation, the proposition is:

"If L be cut in extreme and mean ratio, and there be added to it **a**, then $L + a$ has been cut in extreme and mean ratio, and L is the greater segment."

In other words, if $\dfrac{L}{a} = \Phi$, then $\dfrac{L+a}{L} = \Phi$.

The following table shows the geometric solution.

Step 1: Erect a square on line segment (L = a + b), and form <u>area 1</u> with four areas.
Step 2: Erect a rectangle on line segment **a** and form <u>area 2</u> with three areas.

<u>area 1</u>		<u>area 2</u>	

		a^2	a
a b	b^2	ab	b
a^2	a b	a^2	a
a	b	a	

- <u>area 1:</u> $L^2 = a^2 + b^2 + 2ab$

- <u>area 2:</u> $a*(L+a) = 2a^2 + ab$

Step 3:

We will use proposition 11 in Book II as the template. If <u>area 1</u> equals <u>area 2</u>, then (L + a) is cut in the extreme and mean ratio and (L = a+b) is the greater segment. Therefore, equating <u>area 1</u> and <u>area 2</u>:

$$L^2 = a*(L + a)$$

$$a^2 + b^2 + 2ab = 2a^2 + ab$$

Rearranging: $\qquad a^2 = ab + b^2 = b(a + b) = bL$

Since it was given that L is cut in Φ proportion, <u>area 1</u> equals <u>area 2</u>. Q.E.D.

The following table shows the algebraic solution.

- $\dfrac{L + a}{L} = \dfrac{\dfrac{L}{a} + 1}{\dfrac{L}{a}}$

- Now, it is given that: $\dfrac{L}{a} = \dfrac{a}{b}$

- Therefore: $\dfrac{L + a}{L} = \dfrac{\dfrac{a}{b} + 1}{\dfrac{a}{b}} = \dfrac{a + b}{a} = \dfrac{L}{a}$

3.6 Proposition 6

From *The Elements*: "If a rational straight line be cut in extreme and mean ratio, each of the segments is the irrational straight line called apotome."

Paraphrasing: The length of a given line is a rational number. Cutting this line in Φ proportion produces segments whose lengths are irrational numbers.

Euclid knew that irrational numbers existed, but he couldn't write them down numerically. The decimal point had not been invented. Therefore, he gave them names such as *apotome*.

The proof of this proposition is as follows. In chapter 1.1, we derived the expression:

$$\frac{L}{a} = \frac{a}{b} = \frac{1+\sqrt{5}}{2}$$

This is an irrational number. Therefore, the segments **a** and **b** are irrational numbers.   Q.E.D

3.7 Proposition 7

From *The Elements*: "If three angles of an equilateral pentagon, taken either in order or not in order, be equal, the pentagon will be equiangular."

The proof is shown in chapter 1.5.

3.8 Proposition 8

From *The Elements*: "If, in an equilateral and equiangular pentagon, straight lines subtend two angles taken in order, they cut one another in extreme and mean ratio, and their greater segments are equal to the side of the pentagon."

Paraphrasing: The diagonals of a regular pentagon cut each other in Φ proportion, and the greater segments are equal to the side of the pentagon.

The following table proves this proposition.

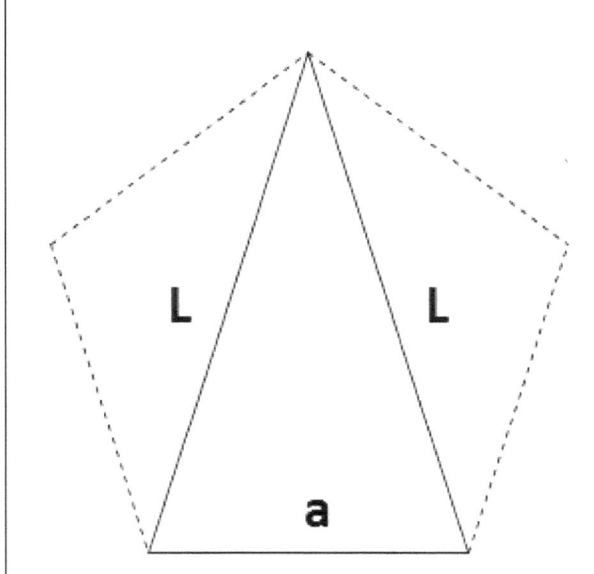

Step 1: When figures 7.a and 7.c are combined, a 36/72/72 triangle is inscribed in a pentagon. Each **L** is a diagonal of the pentagon, and $\dfrac{L}{a} = \Phi$.

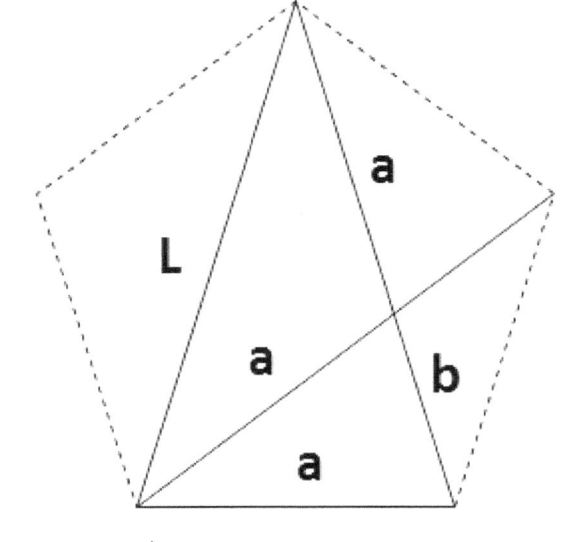

Step 2: A diagonal is added that cuts one of the **L**s into two segments, **a** and **b**, and **a** is the greater segment. By comparing the triangles, **a** is equal to the side of the pentagon. Q.E.D.

3.9 Proposition 9

Quoting *The Elements*: "If the side of the hexagon and that of the decagon inscribed in the same circle be added together, the whole straight line has been cut in extreme and mean ratio, and its greater segment is the side of the hexagon."

Paraphrasing: It is given that a regular hexagon and a regular decagon are inscribed in the same circle. The side of the hexagon added to the side of the decagon is a line that is cut in the extreme and mean ratio, and the side of the hexagon is the greater segment.

The following is the solution. At the top of figure 23, we see a regular decagon, a regular hexagon, and a regular pentagon all inscribed in the same circle with a radius of Φ.

Next in figure 23, we see the regular decagon plotted by itself. From chapter 1.4, we know that a regular decagon has a radius-to-side ratio that is equal to Φ. That makes the side of the regular decagon in figure 23 equal to unity.

Next in figure 23, we see that the side of the regular hexagon is equal to Φ.

Adding these sides together, we get (Φ + 1). This is a line that is cut in extreme and mean ratio, with the greater segment being Φ, which is the side of the regular hexagon.   Q.E.D.

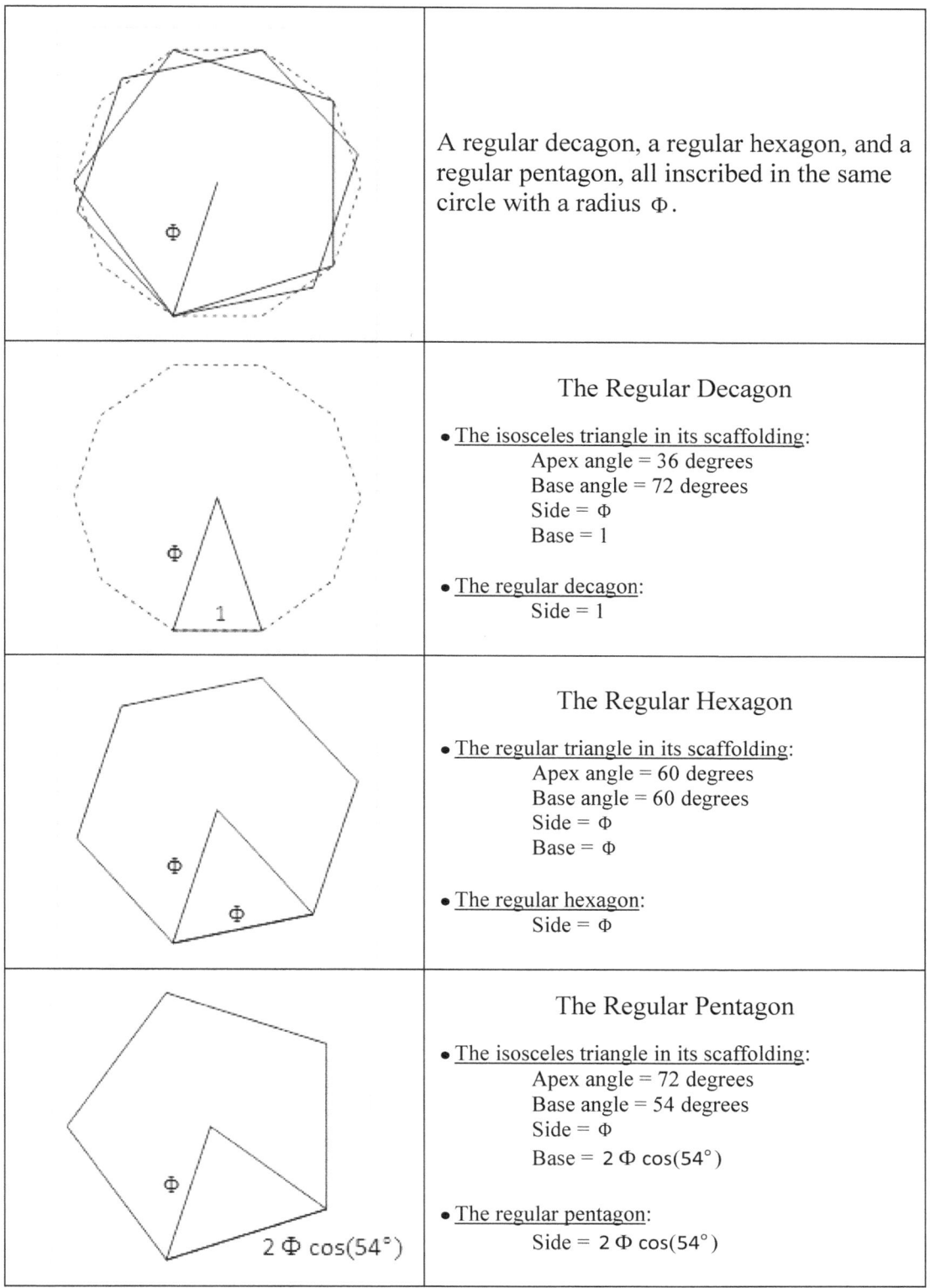

A regular decagon, a regular hexagon, and a regular pentagon, all inscribed in the same circle with a radius Φ.

The Regular Decagon

- <u>The isosceles triangle in its scaffolding:</u>
  - Apex angle = 36 degrees
  - Base angle = 72 degrees
  - Side = Φ
  - Base = 1

- <u>The regular decagon:</u>
  - Side = 1

The Regular Hexagon

- <u>The regular triangle in its scaffolding:</u>
  - Apex angle = 60 degrees
  - Base angle = 60 degrees
  - Side = Φ
  - Base = Φ

- <u>The regular hexagon:</u>
  - Side = Φ

The Regular Pentagon

- <u>The isosceles triangle in its scaffolding:</u>
  - Apex angle = 72 degrees
  - Base angle = 54 degrees
  - Side = Φ
  - Base = $2\,\Phi \cos(54°)$

- <u>The regular pentagon:</u>
  - Side = $2\,\Phi \cos(54°)$

Figure 23: A Regular Decagon, Hexagon, and Pentagon Inscribed in a Circle

3.10 Proposition 10

From *The Elements*: "If an equilateral pentagon be inscribed in a circle, the square on the side of the pentagon is equal to the squares on the side of the hexagon and on that of the decagon inscribed in the same circle."

Paraphrasing: Given a regular decagon, a regular hexagon, and a regular pentagon all inscribed in the same circle, then:

$$\text{(side of the pentagon)}^2 = \text{(side of the hexagon)}^2 + \text{(side of the decagon)}^2$$

At the top of figure 23, a regular pentagon, a regular hexagon, and a regular decagon are all inscribed in the same circle. From this figure:

- side of the pentagon = $2 * \Phi * \cos(54°) = 1.902...$
- side of the hexagon $= \Phi = 1.618...$
- side of the decagon $= 1$

Substituting these into the equation in the proposition:

$$(2 * \Phi * \cos(54°))^2 = \Phi^2 + 1^2$$
$$(1.902...)^2 = (1.618...)^2 + 1 \qquad \text{Q.E.D.}$$

3.11 Proposition 11

Quoting *The Elements*: "If in a circle which has its diameter rational an equilateral pentagon be inscribed, the side of the pentagon is the irrational straight line called minor."

Paraphrasing: We are given a circle whose diameter is a rational number. An equilateral or regular pentagon is inscribed in this circle. The length of a side of this pentagon is an irrational number that Euclid calls *minor*.

The proof of this proposition is given in figure 23. If the radius of the circle is divided by Φ, then the diameter is a rational number. That yields:

$$\text{the side of the pentagon} = 2 * \cos(54°)$$

This is an irrational number. Q.E.D.

3.12 Proposition 12

This proposition does not involve Φ.

3.13, 3.14, 3.15 Propositions 13, 14, and 15

These propositions show the construction of the regular tetrahedron, octahedron, and cube. They do not involve Φ.

3.16 Proposition 16

From *The Elements*: "To construct an icosahedron and comprehend it in a sphere like the aforesaid figures, and to prove that the side of the icosahedron is the irrational straight line called minor."

This figure is drawn in chapter 1.8. It is seen there that its side is the greater segment of a line cut in Φ proportion and hence is an irrational line.

3.17 Proposition 17

From *The Elements*: "To construct a dodecahedron and comprehend it in a sphere like the aforesaid figures, and to prove that the side of the dodecahedron is the irrational straight line called apotome."

This figure is drawn in chapter 1.9. It is seen there that its side is the lesser segment of a line cut in Φ proportion and hence is an irrational line.

3.18 Proposition 18

From *The Elements*: "To set out the sides of the five figures and to compare them with one another."

The five figures are the Platonic solids. After computing and comparing the sides, Euclid is quoted: "I say next that no other figure, besides the said five figures, can be constructed which is contained by equilateral and equiangular figures equal to one another."

Paraphrasing: Euclid proves that the Platonic solids are the only figures that are in the category called regular polyhedron. These are 3-D figures that are bounded by equal straight lines that form equal faces. Euclid's proof is as follows. Let's start with the square.

A Regular Polyhedron with a Square Face

We will start with three squares joined at a point as shown in the following table.

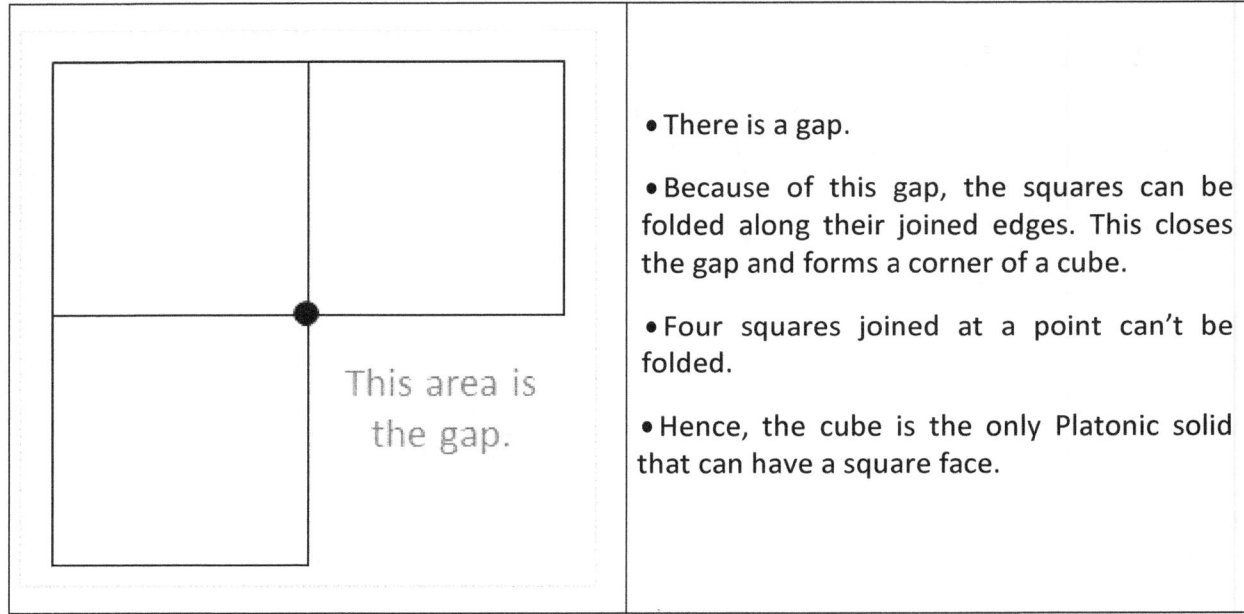

- There is a gap.
- Because of this gap, the squares can be folded along their joined edges. This closes the gap and forms a corner of a cube.
- Four squares joined at a point can't be folded.
- Hence, the cube is the only Platonic solid that can have a square face.

A Regular Polyhedron with a Pentagon Face

The following table discusses using a regular pentagon as a face on a 3-D figure.

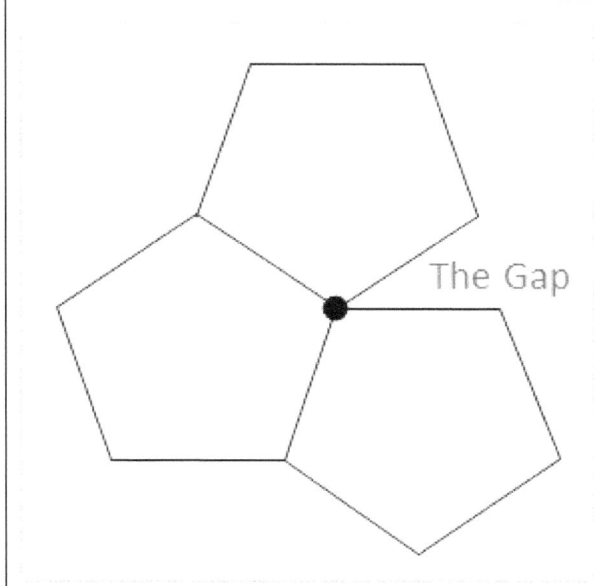

- Start with three pentagons joined at a point.

- There is a gap.

- Because of this gap, the pentagons can be folded on their joined edges. This closes the gap and forms a vertex of a dodecahedron.

- Four pentagons can't be joined at a point.

- Hence, the dodecahedron is the only Platonic solid that can have a pentagon face.

A Regular Polyhedron with a Hexagon Face

The following table discusses using a regular hexagon as a face on a 3-D figure.

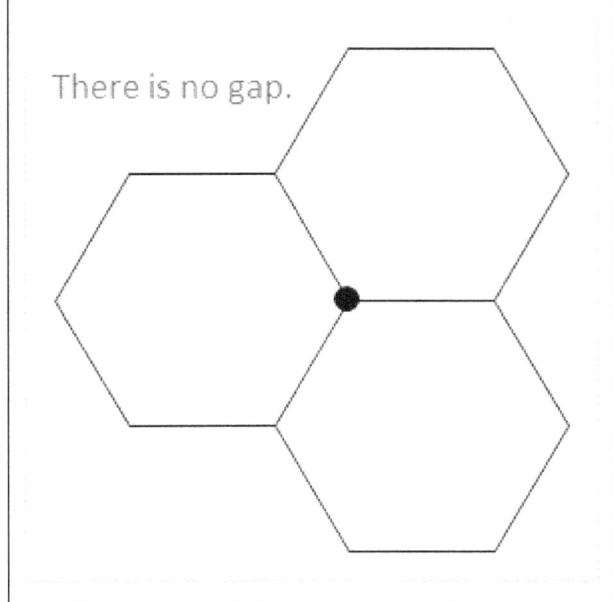

- Three hexagons joined at a point leave no gap. They can't be folded to form a vertex.

- Hence, no Platonic solid can have a face that has six or more sides.

A Regular Polyhedron with a Triangle Face

The following table discusses using a regular triangle as a face on a 3-D figure.

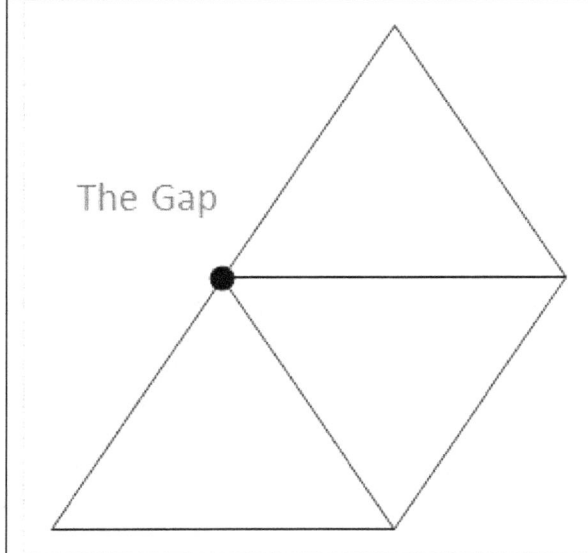

• Start with three regular triangles joined at a point.

• There is a gap.

• The triangles can be folded on their joined edges. This closes the gap and forms a vertex of a tetrahedron.

The following table discusses using four regular triangles joined at a point.

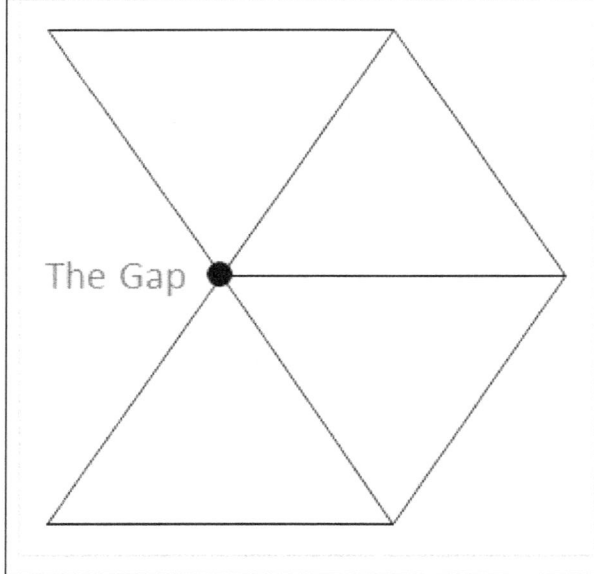

• There is a gap.

• The triangles can be folded on their joined edges. This closes the gap and forms a vertex of an octahedron.

The following table discusses using five regular triangles joined at a point.

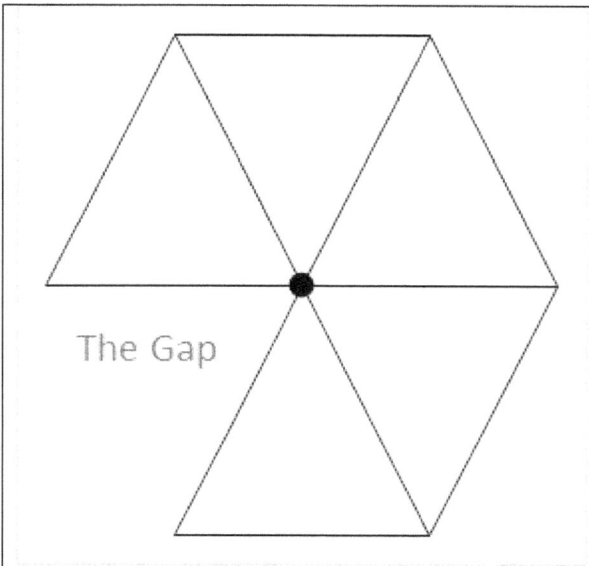

The Gap

- There is a gap.

- The triangles can be folded on their joined edges. This closes the gap and forms a vertex of an icosahedron.

- Adding another triangle would close the gap and there could be no folding.

- Hence, there can only be three Platonic solids that have triangular faces.

We have shown that there can be only five Platonic solids. There is one with a square face, one with a pentagon face, and three with triangle faces.   Q.E.D.

Conclusions

The Elements was a textbook that included all of the math that was known at the time. This included the numbers that were positive integers, fractions, and the "unknowable" irrational numbers.

In figure 3 of chapter 1, we showed how Euclid avoided computing the value of Φ by describing it as a ratio to a known number.

Throughout chapters 1 and 3, we saw how Euclid drew figures using Φ. He makes it a point to show which parts of the figures have lengths that are irrational numbers. The following is a summary:

• The 36/72/72 triangle and the regular decagon are drawn in proposition 10 of Book IV. In proposition 6 of Book XIII, he proves that the lengths of the base of the triangle and the side of the decagon are irrational numbers that he calls an *apotome*.

• The regular pentagon is drawn in proposition 11 of Book IV. In proposition 11 of Book XIII, he proves the length of the side of the pentagon is an irrational number that he calls a *minor*.

• The regular icosahedron is drawn in proposition 16 of Book XIII. In that same proposition, he proves the length of the side of the icosahedron is an irrational number that he calls a *minor*.

• The regular dodecahedron is drawn in proposition 17 of Book XIII. In that same proposition, he proves the length of the side of the dodecahedron is an irrational number that he calls an *apotome*.

In chapter 2, the myths are discussed. Euclid himself says nothing about myths. They arose from very famous historical figures, in times when there were very close connections between religion, science, and math.

There is a common thread among the myths. Φ is an irrational number. In Euclid's time, mathematicians couldn't write down its value using integers and fractions. They had to describe it using words and figures. To many people, the concept of an "unknowable" number was absurd. For centuries, even great scientists and engineers believed that Φ was godlike. At some point, each myth was plausible. But today, irrational numbers are easily handled.

Times have changed. But one thing has not. Already in his time, Euclid was recognized as a great teacher. The following are stories showing that this recognition still exists.

• William Dunham recounts a passage from Carl Sandburg's biography of Abraham Lincoln: "The largely unschooled Lincoln...read Euclid by the light of a candle."[6]

---

6. William Dunham, *Journey through Genius* (New York: John Wiley & Sons, 1990).

• Mario Livio writes that Sherlock Holmes once claimed his deductions were "as infallible as the propositions of Euclid."[7]

• W. W. Rouse Ball[8] retells the following story, which I will paraphrase. A young boy asked Euclid about geometry. "What do I gain by learning all this stuff?" Euclid answered, "Knowledge is worth acquiring for its own sake." Euclid then gives the boy some coppers because "he must make a profit from what he learns."

---

7. Mario Livio, *The Golden Ratio* (New York: Broadway Books, 2002).

8. W. W. Rouse Ball, *A Short Account of the History of Mathematics* (New York: Dover Publications, 1960).

Bibliography

Arasse, Daniel. *Leonardo da Vinci.* Old Saybrook, CT: Konecky and Konecky, 1998.

Ball, W. W. Rouse. *A Short Account of the History of Mathematics.* New York: Dover Publications, 1960.

Bass, Laurie, et al. *Geometry.* Upper Saddle River, NJ: Prentice-Hall, 2004.

Bentley, Peter J. *The Book of Numbers.* Buffalo, NY: Firefly Books, 2008.

Boyer, Carl B. *A History of Mathematics.* New York: John Wiley & Sons, 1991.

Courant, Richard, and H. Robbins. *What Is Mathematics?* New York: Oxford University Press, 1996.

Dunham, William. *Journey through Genius.* New York: John Wiley & Sons, 1990.

Heath, Sir Thomas, trans. *Euclid, the Thirteen Books of the Elements.* New York: Dover Publications, 1956.

Heath, Sir Thomas, trans. *Euclid's Elements.* Santa Fe, NM: Green Lion Press, 2010.

Heath, Sir Thomas, trans. *The Bones, Pocket Guide to the Elements.* Santa Fe, NM: Green Lion Press, 2002.

Hellemans, Alexander, and B. Bunch. *Timetables of Science.* New York: Simon and Schuster, 1988.

Hemenway, Priya. *The Divine Proportion.* New York: Sterling Publishing, 2005.

Herz-Fischler, Roger. *A Mathematical History of the Golden Number.* Mineola, NY: Dover Publications, 1998.

Huntley, H. E. *The Divine Proportion.* New York: Dover Publications, 1970.

Livio, Mario. *The Golden Ratio.* New York: Broadway Books, 2002.

O'Daffer, Phares, and S. Clemens. *Geometry, an Investigative Approach.* New York: Addison-Wesley, 1976.

Olsen, Scott. *The Golden Section, Nature's Greatest Secret.* New York: Walker & Company, 2006.

Posamentier, Alfred, and I. Lehmann. *The Glorious Golden Ratio.* Amherst, NY: Prometheus Books, 2012.

------. *The (Fabulous) Fibonacci Numbers.* Amherst, NY: Prometheus Books, 2007.

Runion, Garth. *The Golden Section.* Palo Alto, CA: Dale Seymour Publications, 1990.

Sazuki, Jeff. *A History of Mathematics.* Upper Saddle River, NJ: Prentice-Hall, 2002.

Simson, Robert, trans. *The Elements of Euclid.* Glasgow: University of Glasgow, 1838.

Smith, D. E. *History of Mathematics.* New York: Dover Publications, 1958.

Sutton, Andrew. *Ruler and Compass.* New York: Walker & Company, 2009.

Sutton, Daud. *Platonic and Archimedian Solids.* New York: Walker & Company, 2002.

Todhunter, I., trans. *The Elements of Euclid for Schools.* London: MacMillan and Co., 1871.

Todhunter, I., and S. Looney, trans. *The Elements of Euclid.* London: MacMillan and Co., 1912.

Vajda, Steven. *Fibonacci and Lucas Numbers, and the Golden Section.* Mineola, NY: Dover Publications, 2008.

Vorob'ev, Nicolai. *Fibonacci Numbers.* Mineola, NY: Dover Publications, 2011.

Walser, Hans. *The Golden Section.* Washington, DC: Mathematical Association of America, 1996.

Westfall, Richard. *Newton and the Scientific Revolution.* Bloomington: Indiana University Lilly Library, 1987.

Appendix: Derivation of the Pythagorean Theorem

The Pythagorean theorem predates Pythagoras. But Proclus, a preeminent Greek historian around 450 AD, incorrectly wrote that Pythagoras was the source. Because of the stature of Proclus, the theorem remains identified as the Pythagorean theorem. Euclid proved the theorem in proposition 47, Book I of *The Elements*. The following is a modern-day proof.

The theorem says that the square of the hypotenuse of a right triangle is equal to the sum of the squares of the other two sides. Consider the following right triangle.

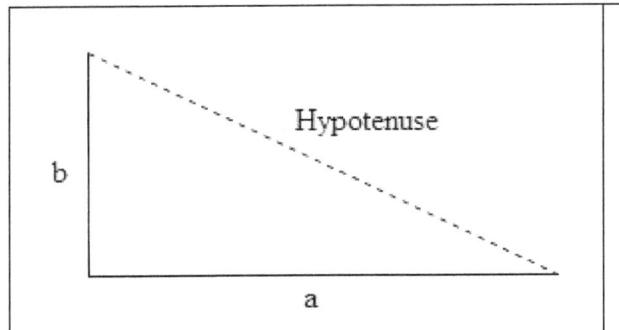

The area of the right triangle is given by:

$$\frac{1}{2}ab$$

Arrange four copies of this triangle, in a square as shown in the following figure.

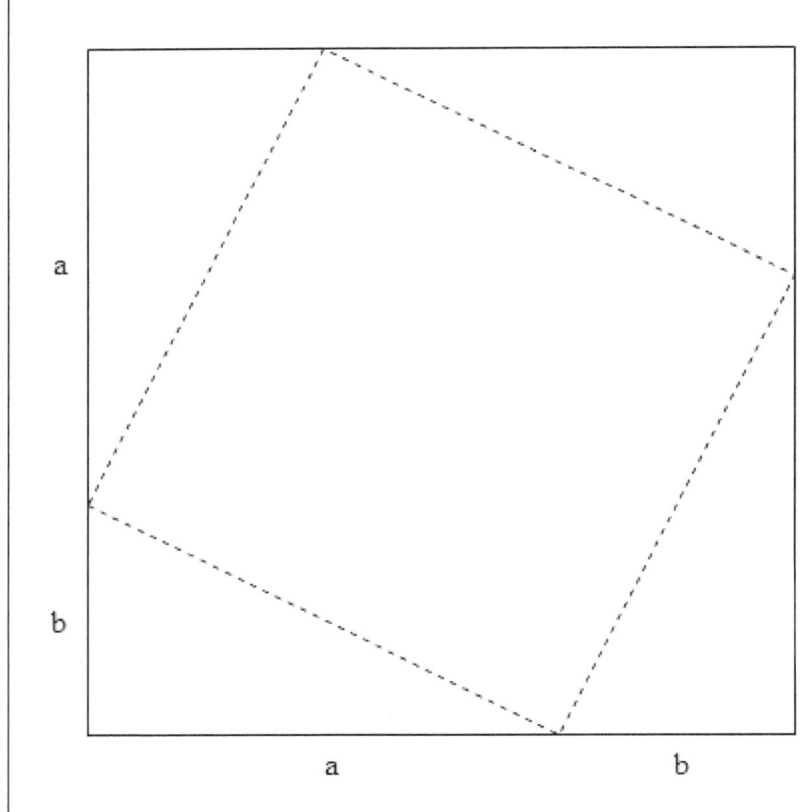

The area of the outside square is $(a + b)^2$. The area of the inside square is the square of the hypotenuse and is what we want to compute.

Denote the area of this inside square as A_H. If we remove or subtract the area of the four right triangles from the area of the outside square, we are left with A_H.

$$(a + b)^2 - 4 * (\frac{1}{2}ab) = A_H$$

Rewriting:

$$A_H = (a + b)^2 - 2ab$$

Since $(a + b)^2 = a^2 + b^2 + 2ab$

$$A_H = a^2 + b^2 \quad \text{Q.E.D.}$$